What people

Small

Surprising, delightful, engrossing, disturbing, and ultimately inspiring, *Small Gods* is a thoughtful and urgent meditation on the ways that life is being re-constituted by technology. Alex Quicho is a fiercely talented young writer who nestles unexpected insights into a well-researched study of contemporary art's engagement with drones, and of the drone's re-configuration of subjectivity and global systems. Relying on extensive interviews and fieldwork, she combines the cathartic poetry of art writing with the cold clarity of geopolitical analysis, and rightfully places these powerful new entities in the long history of a world forged through empire. As a result, *Small Gods* is revelatory.
Vincent Bevins, author of *The Jakarta Method*

Alex Quicho approaches her vivisection of the new, droning flesh with Ballardian playfulness, Mark Fisher's ethical backbone, and Mary Shelley's hunger for new visions. Some books I can see myself reading again and again, always discovering them anew; *Small Gods*, with its commitment to endless frontiers, metamorphosing skies, and slippages of the self and all other borders, is already one of them.
Ales Kot, author of *Days of Hate*

A luminous exploration of drone technology in the gallery and in open air, Alex Quicho gives form to the machinic gaze and asks what we see when we self-surveil, what view of the human is conjured by the drone's-eye perspective. Haunting and revelatory, this book will have you searching the skies above you for the unseen presence of these small gods, their hidden reach.
Alexandra Kleeman, author of *You Too Can Have a Body Like Mine*

Small Gods

Perspectives on the Drone

Small Gods

Perspectives on the Drone

Alex Quicho

Winchester, UK
Washington, USA

JOHN HUNT PUBLISHING

First published by Zero Books, 2021
Zero Books is an imprint of John Hunt Publishing Ltd., No. 3 East St., Alresford,
Hampshire SO24 9EE, UK
office@jhpbooks.com
www.johnhuntpublishing.com
www.zero-books.net

For distributor details and how to order please visit the 'Ordering' section on our website.

Text copyright: Alex Quicho 2020

ISBN: 978 1 78904 004 3
978 1 78904 005 0 (ebook)
Library of Congress Control Number: 2020938311

A CIP catalogue record for this book is available from the British Library.

Design: Stuart Davies

UK: Printed and bound by CPI Group (UK) Ltd, Croydon, CR0 4YY
Printed in North America by CPI GPS partners

Canada Council Conseil des arts
for the Arts   du Canada

We operate a distinctive and ethical publishing philosophy in
all areas of our business, from our global network of authors to
production and worldwide distribution.

Contents

Acknowledgements

Early versions of these essays have appeared in the following publications:
'Sky' in *Real Life* as 'Up and Away', 25 July 2017.
'Falcon' in *Triangle House Review* Issue 8, September 2018.

* * *

This book belongs to Korakrit Arunanondchai, Stephanie Comilang, Anna Mikkola, Lawrence Lek, and WangShui: thank you for letting me into your lives and minds.

My deepest gratitude to Rosanna McLaughlin, for her unparalleled wisdom, sensibility, and editorial eagle-eye.

Thank you to Chomwan Weeraworawit for being the secret orchestrator all the way from Wong's to Pun's. And thank you to Rosalia Namsai-Engchuan, Angela Dimayuga, Abhijan Toto, and all the Ghosts, for opening up new dimensions.

Thank you to the editors who have given my pieces a home: especially to Larissa Pham and Alexandra Molotkow, for their attention to early versions of these chapters, and Michael LaPointe, for the advice and encouragement.

Thank you to Sally Wen Mao, for the words that I returned to so often and for the gracious blessing to open this book with *Live Feed*.

The Canada Council for the Arts provided the resources and funding that made this book possible. I'm grateful for their crucial support of art and artists at a time of unrelenting precarity.

Thank you to my friends for their love, attention, exuberance, and mystique. I'm especially grateful to the brilliant Hatty Nestor for solidarity and motivation; to Natascha Nanji, for Zomia and mind-expansion; to Adrienne Matei, for being my psychic twin and first reader; and to Phil Mak, for putting up with this every day — a superhuman feat.

Thank you to David Crowley, Brian Dillon, Emily LaBarge, and Jeremy Millar for their influence and guidance. Their leadership of the Royal College of Art's Critical Writing programme laid the foundations for this book and the rest of it, along with my CWAD peers, to whom I have bonded for life via squat raves, nude beaches, psychedelic pastorals, and happy fatalism.

Thank you to Wendy Chang for her years of mentorship, and for helping me see beyond the horizon.

Thank you to my family for their unflagging care and belief. And to Ed — always and forever.

Undo

yourself, let the oculus
             burn through my clothes, record

every mistake I make.

Sally Wen Mao, Live Feed

Preface

Not since the atomic bomb has a war technology been so obsessively depicted. Any image of the drone's many forms comes preloaded with dread, whether it's the bulb-headed Global Hawk, able to surveil an area the size of South Korea in a single day; the aptly-named Reaper, which appears as a deadly asterisk when seen from the front, the underside of each smooth wing gripped by a pair of Hellfire missiles; or the comparatively tiny consumer drone, like the DJI Mavic or Phantom, raised eerily into the air by quadcopters. In practice, both military and consumer drones are controlled remotely by human operators on the ground, but 'the drone' is still popularly imagined as a sentient machine, a narrative that is extrapolated from how we first saw and came to understand it: finding and killing people in cold blood, seemingly without anyone behind the joystick or the trigger.

Between 2008 and 2016, the United States intensified its military drone programme, conducting a total of 563 strikes in Pakistan, Somalia, and Yemen. Spun by the US government as laudable acts of 'counterterrorism', the sudden onslaught of drone strikes inspired widespread debate about the circumstances and ethics of drone warfare. While some pundits heralded the strikes as more 'ethical' or 'accurate' than other forms of military action, a first wave of artists — including James Bridle, Trevor Paglen, Laura Poitras, and the collective Forensic Architecture — saw the military drone for what it was: an unprecedented manifestation of asymmetrical power, which enabled the United States government to kill people with impunity, taking violent action in territories it had no declared hostilities with. In an effort to educate and galvanise their viewing public into action, these artists dragged the drone out of military secrecy and into the spotlight, photographing it, making films about it, tracing its shape on city pavements, uploading classified strikes to social media for all to see, often in collaboration with whistleblowers and journalistic organisations such as *The Intercept*. By insistently exposing the military drone and the systems of governance that kept it aloft, they informed our burgeoning understanding of contemporary warfare as constant and uneven, conducted largely in secret, waged through both tight legislation and permissive rhetoric.

The military drone entered public consciousness alongside a broader, existential negotiation that took place over the growing role of technology in our day-to-day existence. We absorbed the violence of its particular context, dropping it into an already potent cocktail of collective fear bubbling just below the slick, celebratory surface of technological innovation. The first consumer drone hit the market in 2010, developed by a French company who gave it the innocuous name of AR Parrot, but it wasn't widely adopted until 2013, when Shenzhen-based DJI released the Phantom: a relatively affordable quadcopter drone that fed back high-definition video and could be controlled through any mobile device. Just as the introduction of the Portapak camcorder in 1967 enabled artists to make movies cheaply without the help of a crew, blowing previous limitations of the filmic medium apart, the DJI Phantom opened up the skies to common use. The aerial perspective was no longer restricted to government, military, and commercial interests, inspiring a fresh coterie of artists to use the consumer drone to explore what it means to see and be seen from above. Recently, exhibitions including the Kemper Art Museum's *To See Without Being Seen* (2017) and the Zeppelin Museum's *Game of Drones* (2019) have attempted to make sense of how 'the drone', as an inconstant entity, so discretely contains our collective fears of surveillance, automation, and asymmetrical power.

Over the past 5 years, the consumer drone's sky-borne form has become stunningly mundane. It tails extreme-sports athletes, makes deliveries, and allows for a genre of selfie that, thankfully, emphasises its subject's insignificance. It's orchestrated into ornamental swarms that bear coloured lights skywards, and is integrated into every pop video requiring a grandiose mansion shot. One would think that the drone, dispersed into a carnival of consumer uses, has tamed its history of violence. About the size of a person's head and even more readily controlled, the consumer drone is easily anthropomorphised, and so somewhat free to take on more benign roles. It has been adopted by artists — such as Korakrit Arunanondchai, Stephanie Comilang, and WangShui — who see something hopeful, or even spiritual, in the consumer drone's ability to decouple vision from the body or send their own perspective far away. It satisfies the age-old ambition to see through other eyes, or to leave one's limited human form behind. In works by Anne Imhof and Anna Mikkola, the consumer drone is also used to invoke a more prosaic feeling: the simultaneous comfort and anxiety of being closely watched. In general, its presence has heightened our awareness of how technologies of tracking and surveillance have become an acceptable part of everyday life. It motivates us to consider why we so desperately want to be watched over, protected, or totally known — a desire that, as we've pursued it, has given machines outsize power over human lives. But though we're most inclined to think of it as an eye-in-the-sky, the drone is usually heard before it is seen. To get why it haunts our collective psyche, we first have to listen to its dread-noise.

01

Dirge

The word 'drone' yawns back to the sixteenth century, when it described both the male worker-bee and the onomatopoeic buzz of its ennui. The 'deep, continuous humming sound' is first organic, rather than mechanical: it would go on to refer to the involuntary, throaty thrum of grief. The drone is linked to the dirge, the twelfth-century Latin mourning song, and comparable to a threnody, the seventeenth-century lamentation that scores a visit from Thanatos, Greek god of death. Consider the first word of the thirteenth-century antiphon for the Office of the Dead, a prayer cycle spoken on behalf of the deceased: *dirige, Domine, Deus meus, in conspectu tuo viam meam*, or *direct, O Lord, my God, my way in your sight*. The dirge pleads for guidance, as does the drone: in its most contemporary usage, a drone is a pilotless aircraft controlled from afar.

The ouroboric roundedness of droning sound contains the quantum ohm and om of nirvana. Drone music, or ambient-drone, is a minimalist music genre that originated in the psychedelic Asiaphilia of 1960s West Coast America, though drone tones are global and ancient, found in the Georgian chant and Japanese gagaku performance. In terms of musical composition, a drone consists of a sustained or repeated note; the tanpura, a long-necked string instrument proportionally similar to a cartoon thermometer, is meant to be played 'unchangingly', its consistent sound providing an active ground for the jump and flow of other instruments. Yves Klein composed his *Monotone-Silence Symphony* in 1949, its first movement a drone, its second silence. The sound was 'drawn out and deprived of its beginning and of its end, creating a feeling of vertigo and of aspiration outside of time', he reflected, writing about the piece a decade later. The intended effect, he said, was to create an embodied experience of noise and silence, as clearly delineated as two blocks of colour in a Rothko painting. Klein's use of sound to achieve a state of temporal transcendence foreshadowed the activities of the American avant-garde. Early American experiments with the drone's abiding intonation invoked ecstatic consciousness, the soothing breathwork of meditation, and a newfound ability to tune out a foreign war. A boy's toy, the drone was solicited by Johns; on its sustenance, John Lennon grew 'Tomorrow Never Knows' (1966) and John Cale 'Heroin' (1967). John Cage, in '4'33"' (1952), made the fullness of the void present. In his audience's own ears arose the drone of their blood, circulating.

Drone music was 'dream music' to experimental composer La Monte Young, a running stream that bore him back to formative idylls. 'The very first sound that I recall hearing was the sound of wind blowing under the eaves and around the log extensions at the corners of the log cabin' in Bern, Idaho, where he was born, he reminisced to an audience in 2015. He was also inspired by the unceasing thrum of power lines and the charged-up whir of electrical transformers, machine-sounds that cycled him out of the woods and into the continuous present. After moving to New York in 1960 to study electronic music at the New School for Social Research, he created the *Dream House* (1962-1990), an immersive environment that took many forms over 4 decades. At one point, it occupied a six-storey building in the New York Mercantile Exchange that Young and his collaborators, including the guru Pran Nath, filled with the hypnotic tones of tanpuras, electronic synthesisers, and a custom-built piano equipped with an extra, lower octave. For as long as the *Dream House* remained in the Mercantile Exchange, the sustained pieces never halted, coaxing listeners — who could wander in and out at will — into deep trances. Entering the *Dream House* dilated one's experience of time; at a diminutive recreation installed at the Guggenheim in 2009, I once lost hours to its mesmerising aura, sitting on the carpeted floor, bathed in magenta light.

How then did this atmosphere of gentle surrender become synonymous with death? In her book *Drone and Apocalypse,* Joanna Demers catalogues the recent history of ambient-drone music with essays on contemporary composers such as William Basinski, Tim Hecker, and the duo Celer, all of whom use a combination of electronic instruments and field recordings to produce sparse tracks of crushing melancholy. Slowing down moments otherwise fleeting or forgotten, stretching the sonic marginalia of daily life into monolithic walls of sound, they demonstrate how the end of the world is already contained in the everyday. 'Apocalypse only magnifies mortality, something already there, on a mass scale,' Demers writes. Taking cues from titles such as Hecker's 'Ravedeath, 1972' (2011) or Celer's 'Elapsed Paradise' (2017), listeners are invited to contemplate an array of morbid themes, from the heat death of the universe to the insignificance of their own existence.

Demers recalls how, at the otherwise raucous 2013 Rock the Garden festival at Minneapolis' Walker Art Centre, the band Low filled their short slot with a single song. Extending their 14-minute opus 'Do You Know How to Waltz?' into a heavy buzzkill, the band bulldozed over its audience's murmurs of dissent, sending amplified echoes in over their heads. At its close, frontman Alan Sparhawk yelped: 'Drone, not drones,' parroting a friend's bumper sticker that, in a post-set interview, he said he found 'fitting' — pitting peaceful practice against killing machine while acknowledging that the two were connected in name. That year, public scrutiny of the US drone programme was peaking. Americans grappled with the cognitive dissonance of how it was that the Obama administration, synonymous with liberal decency, could also be responsible for a sophisticated programme of extrajudicial killing. Luke Heiken, Sparhawk's friend, had created the bumper stickers in response to a recently leaked whitepaper that triggered fresh fury over the drone's unjust conduct.

Here, the drone's double-meaning comes into its own as entrancing noise and wartime weapon. Prolonged exposure to a droning sound can be psychologically affecting. This capacity to alter mental states is what links the absorbing experience of Young's *Dream House* to that of residents of military 'drone zones', where the sound of the drone, which can't be seen from the ground, minces a day into thousands of threatening moments. 'It is a continuous tension, a feeling of continuous uneasiness. We are scared,' said one anonymous source to *The Atlantic* in 2012. 'Drones are always on my mind. It makes it difficult to sleep. They are like a mosquito. Even when you don't see them, you can hear them, you know they are there,' said another.

Though the military drone was put to use by Israel as early as the 1980s, its status as an enforcer of American might spawned in the horror-clouds of 9/11. Banned from non-surveillance uses for decades, the Predator drone was then licensed to kill by the US military, whose troops had been eluded by al-Qaeda and the Taliban, their hydra-headed adversaries. It was reimagined as terrorism's perfect nemesis: the nonhuman hunter of supposedly inhuman motives. 9/11 also spawned *Disintegration Loops* by the ambient composer William Basinski, an hours-long drone piece triggered by his firsthand view of the Twin Towers collapsing. Basinski had been a prolific collector of sound since the late 1970s, recording incidental Americana on magnetic tape: 'the clicking electric buses, the grasshopper legs, the trolleys creaking,' wrote Sasha Frere-Jones of this archive, profiling Basinski in *The New Yorker* in 2014. Unable to afford a Mellotron — a kind of early synthesiser, where each key corresponded to a musical sample recorded on magnetic tape — he created his own by recording snippets of Muzak that pumped through halls, malls, and elevators, 'which [was] like musical anesthesia, the Prozac of the '70s', as he explained in a 2017 interview with *Pitchfork*. 'When you'd slow it down and look at it under a microscope, there's this well of melancholy down there. So I started getting into the idea of pulling things out of thin air, making something from nothing.'

Basinski stored his collection badly, letting some of the loose tape dangle from a tree branch he'd bolted above his desk, like entrails spoiling in the sun. At the time, he had been experimenting with the damaged recordings, feeding them through his stereo at full doomsday volume, listening closely to how decay had changed them. The morning of 9/11, he was up early to prep for an interview at Creative Time, an arts organisation based at the World Trade Center, when he first got wind of the attacks. He climbed up to his rooftop for a better view of the towers, watching the smoke cloud rising over the city. Instinctively, he reached for his camcorder to film it. When he put the footage on loop in the studio later that day, curious about how its presence could affect his nascent compositions, 'something clicked', wrote Frere-Jones. Basinski spun his tapes into an elegy. The resulting *Disintegration Loops* are weird, wobbly, stuck in recursion. Noise unspools into loose drifts, the echo of collapse in slow motion. Repetition becomes eerie with an afterimage of plummeting trauma, of the rising cloud of debris that undoubtedly contained human ash. As Basinski drew a new, uncertain form out of destruction, so too did the United States, re-emerging from the ashes of Ground Zero as a paranoid totality, ready to wage war on 'terror'.

02

Spectrum

Picture a declassified drone strike video. The ticking indicators of time, altitude, and location blocked out by grey rectangles that frill the screen ornamentally, with only a floating N indicating the cardinal direction. Crosshairs mark its centre. The landscape rotates around this miniscule axis, swaying like a drunk compass needle, slowly fixing on the brushed grey outline of a building roof. The camera swallows a dirt road, and another building hunched alongside, before the strike flashes down. A total, seared white — until billowing smoke clouds emerge from interiors flared with light. Fire consumes the target, igniting the building's outline against the pocked and tamped earth. In military literature, drone vision is described as 'multispectral', made up of both visible and invisible worlds. Come nighttime, pilots switch to infrared, using thermal imaging to see bodies in the darkness. Drone cameras pick up on the radiation that emanates from warm-blooded beings, lighting them up against the comparatively cold landscape. These signs of life begin a person's transformation into a ghost.

'When the smoke cleared, the man who had run to the other two was rolling on the ground, clutching his leg in desperation, and I watched his life's blood spurt out in the rhythm of his heart,' wrote whistleblower and former United States Airforce sensor operator Brandon Bryant in a now well-circulated letter, describing his first kill — and in the process, sharing details of bloodshed that are typically kept from view. 'It was January in the mountains of Afghanistan. The blood cooled. He stopped moving, eventually losing enough body heat to become indistinguishable from the ground on which he died.'

We tend to view the drone as occupying one of two poles:

either a high-definition precision instrument (weapon, or camera, or both), or else as a redacted entity, a proxy soldier hidden from public view by military classification. Both are complex deceptions, sustained by the machinations behind state violence. The drone is better understood somewhere in the middle of a spectrum, where survivors' memories render the strikes at a human scale. While useful as a way to expose the military operations of a clandestine war, the testimony of drone pilots does little to foreground their victims' lives, reproducing only the high drama of the strike — and perhaps misplacing sympathy with the men and women who sit in their air-conditioned shipping containers, thousands of miles away from the action, sifting enemies out of the moving figures on-screen. Drone survivors speak in the real world, though their presence may be undermined by an array of political forces, their memories fragmented by post-traumatic stress. The insistence that the drone is all-powerful yet invisible, infallible yet opaque, depends on the erasure of their accounts.

* * *

Between 2013 and 2014, Forensic Architecture — an agency comprised of artists, architects, and filmmakers which investigates cases of human rights violations with, or on behalf of, affected communities — conducted investigations into drone strikes that occurred in Pakistan, Afghanistan, Yemen, Somalia, and Gaza from 2004 onwards. Through gathering survivor testimony and conducting detailed analyses of the architectural ruins left by drone strikes, the collective's work aims to counteract the state-enforced systems of invisibility that have enabled these secret wars to take place.[1] 'In a way failure is our starting point,' says Christina Varvia, deputy director of Forensic Architecture, in a 2018 interview with *Dezeen*. 'We take cases that might never lead to anywhere, but we still think that it is important to be talking about those things and archiving those events by way of analysis and research. This is also a form of resistance. When there is an effort to erase those violent events, just the effort to record, to clarify and to debunk the official line, it's an act of resistance towards that erasure.'

In his text, 'Violence and the Threshold of Detectability', Eyal Weizman, architectural theorist and founding member of Forensic Architecture, describes how the 'material limit of images', from the size of film grain to the resolution of publicly-available satellite views, is used to obscure or contradict violent acts. In the year 2000, he writes, Holocaust denier David Irving attempted to sue an American writer for libel, contesting her accusation that he was 'falsifying history'. The nature of the case demanded, in essence, a re-assessment of historical evidence, and centred around a series of aerial photographs taken of Auschwitz by the US military during World War II. To Irving, a photograph of people being marched through the death camp did not depict human beings; instead, he claimed that the human heads, as seen from above, were nothing but 'brush strokes' dappled on the negative's surface, a clear indication that the image had been tampered with. Nevin Bryant, a NASA aerial and satellite photography analyst, presented a detailed counter-statement that swiftly disposed of Irving's claim, but not without noting that the distortion that Irving attempted to use as counter-evidence originated in a weakness of the photographic medium. From the height that each picture was taken, a human head was roughly equivalent to a single particle of silver halide, the light-sensitive chemical compound that is mixed with gelatin to create photographic film — creating the physical 'moire-effect' that destabilised, for a moment, the photograph's status as undoubtable evidence.

Comparing the silver crystal and the pixel, Weizman contrasts the photograph's material condition of opacity with its metaphysical promise of clarity. In popular understanding, drones' deadly efficacy has become synonymous with the machine's power of sight; they're spoken of as all-seeing, unblinking eyes in the sky, regardless of their limitations in practice. The rhetoric of the drone as a high-definition observational tool is used by those in favour of extrajudicial killing, too, as if to suggest that attentive pilots could read a target's guilt or innocence as easily as they do the drone's location or cruising altitude. In reality, however, the drone has relied on an infrastructure of poor images to conceal and continue its attacks. According to US regulations effective until 2014, publicly-available satellite images were degraded to a resolution of 50 centimetres squared, equating one pixel to the approximate dimensions of a human body as seen from above. The low resolution protected individual privacy while concealing strategic locations and acts of violence such as drone strikes. Defined by Weizman as the 'threshold of detectability', this dimensional specificity ensured that scattered corpses, or the signature of a Hellfire missile as it passed into a building, remained practically unseen, appearing as a mere smudge on a rooftop or a darkened patch on the ground. By exerting legal control over images, the US government ensured that classified operations remained as such — allowing the CIA to claim that it could 'neither confirm nor deny the existence or nonexistence of such targeted assassinations'. As of 2017, however, the 'American Space Commerce Free Enterprise Act' has lifted legal restrictions on the resolution of commercially-available imagery — instead relying on both censorship and the buying of exclusive rights to certain images to ensure that classified areas remain invisible.

Visual opacity is but one method of state control, working in tandem with other, more overt tactics aimed at disappearing evidence and silencing the voices of witnesses — such as the Pakistan government's media blackout of the former Federally Administered Tribal Areas (FATA), a once semi-autonomous region of Pakistan that was brought under full government control in 2018. In the early 2010s, when Forensic Architecture first began to investigate drone strikes in the FATA, the region had been a conflict zone for well over a decade. Allied with the United States since 2001, Pakistan had been conducting back-to-back US-supported military operations in the area, targeting al-Qaeda and other armed militant groups. To control how its controversial actions were represented, the government of Pakistan curtailed how information flowed out of the FATA, intermittently shutting down all internet services and enforcing military checkpoints along its borders to ensure that recording devices did not enter or leave the area.

Media bias has further strengthened the drone war's impenetrability. A case in point: The *Drone Strikes Platform* (2014), an interactive database of US drone strikes in Pakistan created by Forensic Architecture and the Bureau of Investigative Journalism, found that 61 per cent of US drone strikes in Pakistan had targeted civilian homes. Yet the Western media has largely elected to call these 'compounds', estranging them from any associations with the domestic to invoke something more communal, walled, and organised; in short, a place that houses combatant groups rather than families. Four in-depth investigations — *Drone Strike on a Jirga in Datta Khel* (2013), *Drone Strike in Mir Ali* (2014), *Drone Strike in Miranshah* (2014), and *Knock on the Roof: Drone Strike in Beit Lahiya* (2014) — attempt to reveal the reality, piecing together photographs, videos, and survivor testimonials to provide evidence of the drone wars' toll on civilian life.

Commissioned by the UN Special Rapporteur on Counter Terrorism and Human Rights, *Drone Strike in Mir Ali* is presented to the public as a short video which summarises the rigorous investigative process and key findings. Here, an anonymous source reconstructs the drone strike that she survived with the help of a 3D modeller, who builds digital doors, windows, props, and textures, placing them around the model according to her instructions. Together, they produce a mud-walled house, sparsely furnished with iron-framed beds, cooking implements, and the colourful, scattered Lego blocks that once belonged to the survivor's child. She lingers over the exact position of a door, picking out the correct floor coverings and hanging lines of phantom laundry. The 3D model becomes a 'memory palace' — a mnemonic tool first cited in Cicero's 55BCE dialogue, *De Oratore*. In order to recall a vast amount of information, or memorise a long and complex speech, an aspiring orator only has to visualise a space — a home or a palace — and arrange their knowledge within it like so much furniture. Then, the theory goes, they could visit their mind's domicile and find the material they need, right where they left it.

Forensic Architecture's methods extrapolate from Cicero's ancient tactic, building a 'memory palace' with the victim to aid the production of a more complete testimony. Details are crucial to corroborating evidence, yet survivors of violent acts commonly find that their recollection is uneven, the scrim of memory chewed and punctured by what transpired. A traumatic event will often submerge the possibility of its own retelling. The brain cauterises its own wound. 'Such testimonies are, as a result, rarely straightforward records of events, and should not be interpreted only by what they contain,' write the Forensic Architecture team. 'Rather, what is missing, distorted, or obscured becomes equally instructive.'

As she pieces together her former home, one recollection leads to another; soon, the structure is complete enough to run through the day's events. The modelling technician animates the space, switching the program's perspective to human eye-level. The anonymous woman describes how the strike shook debris loose from the earthen ceiling: how at first she had to feel her way through thick smoke before she emerged outdoors to find body parts and debris strewn across the property. The ground around the impact crater was oddly patterned from the explosion: distinctly striped, rather than completely charred. She remembers how her husband had discovered a severed hand among the rubble, and how, caught in the blades of the electric fan — which she had earlier remembered to place in the virtual courtyard — were pieces of hair and flesh. The scattered, miniscule remains were painstakingly collected, she mentions, for the funerals.

* * *

One of the drone's most pervasive myths is its precision. With a kill radius of 13 to 20 metres, it does not dramatically minimise collateral damage — in contrast, a haphazardly tossed grenade has a 3-metre impact zone— though such a line has been used to promote death by drone as somehow more ethical than by any other weapon. Drones have taken the lives of thousands — though how many, exactly, continues to be contested. In Afghanistan, Pakistan, Yemen, and Somalia, the Bureau of Investigative Journalism estimates nearly 10,000 people have died in 4413 US strikes since 2002, with over 1400 of those classed as civilians. A total of 331 were children. In a ledger called *Naming the Dead* — maintained by the Bureau of Investigative Journalism, which sifts names, and sometimes faces, from the aftermath of a drone strike — are Daoud and Murtaza, slain in Afghanistan aged 3 and 4. There is Umm al Shaya, shot down with her children in Waziristan, and Bibi Mamana, murdered as she picked okra outside her family home. And there is Tariq Aziz, targeted for speaking out against the torture of the drone hovering above his home.

'A lot of the kids in this area wake up from sleeping because of nightmares from them and some now have mental problems. They turned our area into hell and continuous horror, day and night, we even dream of them in our sleep,' 13-year-old

Mohammed Tuaiman told *The Guardian* just weeks before he was slain in Yemen. 'I saw all the bodies completely burned, like charcoal,' said Tuaiman's brother after the boy's death. 'When we arrived we couldn't do anything. We couldn't move the bodies so we just buried them there, near the car.'

Using 22 seconds of video of the aftermath of a US drone strike in the FATA, painstakingly smuggled out of the region, the Forensic Architecture researchers were able to reconstruct the event in *The Architecture of Hellfire Romeo: Drone Strike in Miranshah, Pakistan* (2014), proving that the blast had, in fact, killed four people. The two videos were pulled apart into split-second stills, not unlike how a butcher — or an autopsist — will break down a carcass into an array of newly whole parts. In the finest slice of the visual plane, what may have been otherwise overlooked becomes clear as day. In their exhibition *Counter Investigations* at London's Institute of Contemporary Arts (2018) the collective describes how 'the "split-second" is the default temporality of forensics'. It is the atom, if you will, of perception: the smallest unit that still retains the essence of the whole that it composes.

The stills were then mapped across a 3D rendering of the destroyed home. 'The walls are like photographs, exposed to the blast like film is exposed to light,' says the video's narrator, describing a rigidly rendered diagram, which shows where pockmarks of shrapnel, viscera, and pulverised human bone were etched forcefully into the walls by the blast. At the heart of the morbid nebula, two human silhouettes appear in negative, their bodies having absorbed the exploding matter. Recorded in the architecture itself, the reverse image becomes deployable evidence counteracting CIA efforts to disguise and deny the strike.

Noting how the missile passed cleanly through the building's roof and upper floor before detonating, Forensic Architecture attributed the blast to the Hellfire II (Romeo), a delayed-fuse missile launched by an American drone. Developed by US aerospace defence firm Lockheed Martin as a lucrative improvement to the original Hellfire, the Romeo's key update was — once again — precision, specifically the ability to pass cleanly through roofing and detonate only once it's in the same room as its target. It is worth noting that the military argue such developments to weaponry make the strikes 'humanitarian', because the missiles are designed to detonate from within, leaving a building's outer structure intact and therefore minimising damage to other people or property. Such arguments leave out the fact the Hellfire II has simply enabled armed forces to better target urban domestic spaces, and better obscure the evidence after the fact.

* * *

Where else could the drone be precise? In how it studies and selects its targets? Even as its pilots cruise slow over family 'compounds', watching for the errant movement that will turn an individual into a mark, claiming a drone's accuracy is 'tantamount to saying that the guillotine, because of the precision of its blade...makes it thereby better able to distinguish between the guilty and the innocent,' as Gregoire Chamayou, author of *Drone Theory* (2013), writes. A single drone can linger, fully armed, above a potential target's home for up to 14 hours. Once it runs low on fuel, it can be replaced by another — allowing a video feed analyst to spend days, or even months, on end observing the lives below, keeping pilots updated via instant messaging. In US military language, a 'surgical strike' sees drone pilots zoom in close and act quickly, while the 'signature strike' uses previously collected information, such as metadata, to determine a known target's guilt or innocence. Described in technical terms, the process of selecting a target could appear as thoroughly forensic as a police investigation, but watching for the detail that will condemn a person to Hellfire is an inexact science, depending on whether the analyst can tell a wedding from a briefing of militants or a child from a soldier from 30,000 feet in the air. Many drone operatives are convinced of the technology's accuracy. 'I sometimes say we're fighting with laser beams against people who beat each other to death with sticks,' says an officer identified as 'Maj. Jeffrey' in an article by *Guardian* journalist Roy Wenzl, reporting from an air base in Kansas. Yet in practice their actions are rife with uncertainty. As Wenzl writes, the most frequent phrase uttered in a drone command centre is 'Don't push the button', as analysts second-guess their own judgement. Following a strike that mistakenly killed 25 Afghan civilians including two toddlers, a sensor operator, two feed analysts, and a drone pilot sheepishly discuss the mistake that cost so many lives:

04:40 GMT (9:10 am Afghanistan) Predator crew begins spotting more women and children
04:40 (Sensor): What are those? They were in the middle vehicle.
04:40 (MC): Women and children.
04:40 (Sensor): Looks like a kid.
04:40 (Safety Observer): Yeah. The one waving the flag.
04:40 (Bam Bam41): And Kirk97, Bam Bam41.
04:40 (Pilot): Bam Bam41 go for Kirk97.
04:40 (Bam Bam41): Roger, Kirk97. Do you have any updates on what you're able to see at the engagement site?
04:41 (Pilot): Kirk97. Uh, negative, we are still observing at this time. Still no weapons PID, everything else matches with your assessment. Uh, still looking.

As Forensic Architecture's investigation of the *Drone Strike in Beit Lahiya* reveals, the way in which drone wars are administrated insulates their executors from questions of guilt or innocence.

Before sending missiles into civilian areas in Palestine, Israel's armed forces first alert inhabitants via text message, telephone calls, air-dropped pamphlets, or the firing of small warning shots referred to as 'a knock on the roof'. A short window of time, typically between 10 to 15 minutes, is allowed for residents to evacuate (to where, one wonders); anyone choosing to remain in their homes thereafter is designated a 'voluntary human shield' in the newly-declared combat zone. As writ into international law, such a declaration effectively protects the army from accountability, as the survivors of a warning strike, which killed two women and four children, discovered when they attempted to sue the Israeli state. 'Maybe it was [the pilot's] mistake,' said the judge, as recounted by the survivors' lawyer. 'But the state is not responsible for killings resulting from an act of war.'

Though *Drone Strike in Beit Lahiya* did little to sway the Israeli courts' final decision, it provided crucial evidence that the drone wars aren't an accidental intrusion on civilian life, but rather that their conduct is painstakingly legalised and tightly administered in order to liberate violence from its agreed constraints. The protection of 'civilian territory' no longer holds — the declaration of a site of drone war happens anew with each target that is approved for extrajudicial killing. 'The body becomes the battlefield,' writes Chamayou in *Drone Theory*. The distinction between bystander and participant dissolves, as 'combatant' now means any military-aged man who happens to be in a conflict zone — the boundaries of which are always in flux, encompassing single homes or entire regions.

03

Sky

'Bed down locations' — spy slang meaning where an enemy sleeps — are littered across the Afghanistan badlands, where the arid soil is replete with wildflowers. The region's steppes and valleys have acted as sites of refuge for millennia, shielding species from ice age cycles which would wipe other regions, like the British Isles, clean of life. Frustrated by geologic irregularity, British invaders in the 1860s nicknamed the bordering region in Pakistan 'Hell's Door Knocker'. Today, the landscape continues to hide humans fleeing cross-continent invaders. In order to avoid heat-seeking drones, people drench themselves in cold water, throw plastic tarps over their bodies, or burrow into earth-made hiding places.

At Laura Poitras' *Bed Down Location* (2016), part of the documentary filmmaker's 2016 exhibition *Astro Noise* at the Whitney Museum of American Art, the potential target of the drone is the viewer. There, I clamber onto a knee-height platform and recline beneath a screen suspended above to watch a soothing chroma-shift from day to night. Skies from over Somalia, Yemen, Pakistan, and Nevada cycle through projections; soon, I'm lulled into watching for the streak of a drone crossing, like a shooting-star, overhead.

Poitras renders skyscapes as HD dreams: high cerulean, pink and white, velvet black. The visitor is meant to form an affinity with those living under these heavenly slices, the familiar constellations a reminder of others fated to death on the wrong side of asymmetrical wars. As I look up I let my mind drift, and associations float light as clouds: red morning, sailor's warning; *blue skies smiling at me*...The endless clarity of these skies begins to feel unsettling. Soon, the crackling, faraway sound of the drone pilots' communication fills the air around us in the museum, and I remember 13-year-old Zubair Rehman's testimony at a congressional hearing in Washington, DC 2013: 'I no longer love blue skies. In fact, I now prefer grey skies. The drones do not fly when the skies are grey.'

In *Bed Down Location*, the visitor is offered a flipped view of drone target sites, reversing the sightline of American power relations. Poitras' work allows viewers to see these areas from the ground up, countering how we typically view them via satellite feed, where dusty roads and the outsized pixels of buildings make up the barrenness of a bird's eye view, if we see them at all. I think of Charles and Ray Eames' film *Powers of Ten*: how alarmingly quickly, as the camera pulled ever farther from Earth, the picnicking couple was swallowed up by the planet. Poitras makes one thing clear: regardless of one's own agency, those on the ground are at the very bottom of a vertical hierarchy of power, a speck observed by a system that sees — and therefore controls — all.

'Would you feel any pity if one of those dots stopped moving forever?' says Harry Lime, played by Orson Welles in the post-war noir film *The Third Man* (1949). His speech is flat, gangster-affected; he's looking down into the murk of a carnival from a Ferris wheel's revolving zenith, seeing only ants in the people below. 'If I offered you £20,000 for every dot that stopped — would you really, old man, tell me to keep my money? Or would you calculate how many dots you could afford to spare?' Lime, suspended just a hundred feet above ground level in a glass cabin, easily pictures bullets rending the crowd. Fifty years early, he makes a case for vertical vision's disembodying effect. His argument is echoed by those who see drone warfare as a 'push-button war' that turns murder into a numbing game, placing the killer's eyes in the sky and their body thousands of miles away from their victims. Drone pilots cruising over Pakistan's Khyber Pakhtunkhwa region would, however, see faces rather than dots. Unfurled across rooftops and laid out on fields, high contrast and massively scaled, poster portraits of child-victims look back up at the sky. Titled *#notabugsplat*, the posters are the work of an anonymous Pakistani art collective who installed these images as a soft defence. The project's title counters US military slang which equates victim with 'bug splat', also meaning the irradiated spread of collateral damage on a tactical map.

Containing the perspectives of drone pilot and victim, *Bed Down Location* wouldn't look out of place in a natural history museum of the distant future: *world's-eye-view, circa 2016*. Its soft lighting, reminiscent of the light-cycles used on transoceanic flights to ease jet-lag upon arrival, also invokes psychic transit; *Astro Noise's* curator, Jay Sanders, likens it to the sublime sky-slices of James Turrell. The ambiguity of experiencing *Bed Down Location* allows the drone to represent more than just its violence. Within the drone's body coheres military and industry, surveilled and surveillor, capitalist superpowers and disenfranchised populations.

* * *

Bodies emit self-betraying frequencies, streaming pheromones, radiant heat, cellular signals, the minute tick of a slowing pulse. Unknown to viewers, while they're beneath the screen of Poitras's *Bed Down Location*, their bio- and metadata is gathered and broadcast into the exhibition's final room. Side-by-side screens display the heat signatures of viewers lying on the platform just a moment ago, and details gathered from every mobile phone that has passed through the exhibition space. It's a re-enactment of a real National Security Agency tactic where, according to investigative journalist Jeremy Scahill in his book *The Assassination Complex* (2016), surveillance pods fixed to the bottom of aircraft 'vacuum up massive amounts' of metadata from entire towns in order to sift out their targets. Replicating the relationship between watcher, watched, and command centre, the piece emphasises the creepy mundanity of being observed from a distance.

Surveillance is a prominent theme in Poitras's work: her best-known documentary, *CITIZENFOUR* (2014), took US whistleblower Edward Snowden for its subject. The Whitney exhibition *Astro Noise* was named after a communiqué from Snowden to Poitras, sent in 2013: an encrypted file of the same name containing evidence of the National Security Agency's mass surveillance of the US public. With the help of Poitras and Guardian columnist Glenn Greenwald, Snowden unveiled the extent of global surveillance, at that point yet unknown. As unruly and misattributed as hysteria, the American fear of terrorism was used to justify sweeping privacy violations. *CITIZENFOUR* is the final instalment in Poitras' 9/11 Trilogy, which also includes *My Country, My Country* (2004) and *The Oath* (2010). *My Country, My Country* follows prominent doctor Riyadh al-Adhadh as he lives and works in Baghdad during the Iraq war. It also contains the 8-minute scene that eventually landed Poitras on the US government's terrorist watchlist: after she caught a fatal skirmish between US soldiers and a group of insurgents on film, some of the Americans accused her of knowing about the attack in advance. 'You have a threat score that is off the Richter scale. You are at 400 out of 400,' one border guard told her. Poitras suspects that it was their accusations, however unfounded, that got her name on the watchlist, in turn leading to countless detentions and interrogations. Altogether, the trilogy of films enlace the nexuses of the War on Terror, moving from Iraq to Afghanistan, Guantanamo Bay Prison, the US Supreme Court, and eventually Snowden's own bedroom, in order to expose, however unevenly, some of its sprawling machinery. Though Snowden's leak marked a paradigm shift in how Americans conceptualise privacy, the NSA's transgressions were quickly accepted as a new status quo. Coinciding with the rise of social media, personal exposure began to feel as habitual, and apparently essential, as linking one's hands above one's head for an airport body scan. (When these same scanners were first introduced in 2007, they caused a scandal, drawing accusations of indecent exposure.)

Through the arrangement of *Bed Down Location's* elements, Poitras invokes the duplicitous nature of the screen. Looking up at its false skies, most viewers assume that their experience is passive and one-sided. They are shielded from the knowledge that their every move is actively sensed by the monitor, which they assume only exists for their pleasure. Completing Poitras' diorama of drone warfare, the reveal in the next room also mimics the moment of public awakening to mass surveillance. *Bed Down Location* is an unsubtle prompt to consider how our screens enable corporations and the state to become intimate with our speech and habits, narrowing the existential gulf between American gallery-goers and residents of the drone zone by prompting the former to consider how such powers of observation can be instrumentalised. Less intentionally, *Bed Down Location* reinforces how an American audience is mostly ambivalent to these dangers. Laid out on the slab in the museum, viewers look like they're awaiting alien abduction, not earthly destruction. Some are resting after a day spent traversing New York City, untouched by the paranoia the exhibition aims to evoke.

Under the screen, I think of a different set of celestial visions: photographs taken by Trevor Paglen, which I had seen at the Institute of Contemporary Art in Boston some years prior. From a distance, Paglen's photographs of the vast Nevada sky could seem as ordinary as stock photos — but hidden in plain sight, crossing swathes of candy-blue and sherbet-orange, Reaper and Predator drones take test-flights through the restricted airspace. With narrow wings and domed heads that resemble those of beluga whales, these particular drones have a distinct,

Sky

if ungraceful, silhouette. Paglen's *Untitled (Drone)* (2010-11) series captures none of this detail; to a careful observer, the drones are grey smudges in the multicoloured sky-show, an almost molecular pattern of diffuse reflection, a fleck in the film grain. For the most part, the sky looks empty — it seems there's nothing there at all.

'The drone is a vision machine that is intended to remain invisible,' write curators Svea Bräunert and Meredith Malone in their 2016 essay 'Bringing the War Home'. How it eludes direct observation is not solely a consequence of how it flies. Much like drone strikes themselves, the drone's status of invisibility is strictly enforced by government bodies, military protocol, and international policy. As Paglen has written, secrecy 'nourishes the worst excesses of power', creating a permissive culture around the abuse of human rights. In such a climate, it follows that exposure is considered by many as the only recourse to corrective action; hence the urgency of whistleblowers and the pursuit of investigative exposés, all amplified by concerned citizens through social channels.

Yet *Untitled (Drone)* is not a collection of documentary photographs, snapped to bring dark secrets into the light of day. Smeary, out of focus, and, in his own words, 'useless as evidence', they capture the limits of vision: where sight breaks down and degrades. Using lenses designed for astronomical photography, Paglen strains the camera's eye towards classified sites from the closest spot on public land. 'In some ways, it is easier to photograph the depths of the solar system than it is to photograph the recesses of the military industrial complex,' he has written, words which may not be as hyperbolic as they first seem. For the series *Limit Telephotography* (2012), he photographed military 'black sites' from miles away, allowing the desert air to turn his images into phantasms. 'Between Earth and Jupiter (five hundred million miles away), for example, there are about five miles of thick, breathable atmosphere. In contrast, there are upwards of 40 miles of thick atmosphere between an observer and the sites depicted in this series,' he writes, comparing the difficulty of making *Limit Telephotography* to the relative ease of an older work, *The Other Night Sky* (2010-11), where classified satellites appear as starry, time-lapsed smears above the US landscape.

A difficult process has not created difficult images. They are, for the most part, dreamily arresting: 'miragelike photographs of visions collapsing, of light falling apart.' Their beauty, Paglen says, is a formal strategy, enticing a viewer to explore the complexity he opens up through sidelong streams of writing, teaching, and lecturing — all of which he sees as part of, not supplementary to, his artistic practice. He restrains our desire for immediately interpretable images, which he associated at the time with systems of observation and control.

* * *

In an interview with Poitras, artist and writer Hito Steyerl speaks about how 'the drone shadow becomes a sort of university'. Conscious of the drone pilot's train of thought, people avoid any sudden movements, sport, or celebration. Instead, they talk in low voices, listen to the radio, write, and read books to pass the time. 'In war, you grow accustomed to adopting new routines, creating entirely new traditions,' writes Atef Abu Saif in his account of the 2014 invasion of Gaza, *The Drone Eats With Me*. 'Instead of living in a labyrinth of the unexpected, you impose repetition, make every day a duplicate of the last.' One balmy night, Saif's friend cautions him not to smoke shisha outside, lest the drone pilot mistake the glowing coal for a weapon. An unintentional side-effect of caution is downtime, and the closest they get to a party is a barbecue for the birthday of Saif's son, held in a stairwell so that its heat signature is concealed from the drone's sight.

Beyond each day's work of writing a column for the *Al-Ayyam* newspaper, Saif's main semblance of order is tethered to his friends. He'll go to Wafi's house, 'too narrow for the patience of the drone operator to target', to talk over shisha, then to Faraj's for its robust electricity — a rarity during the war — and big-screen TV. Finally, he'll turn up at Hussain's to snack on Egyptian delicacies, blessed by the gestalt of his friend's homesickness.

To Saif, the drone is an unpredictable force, shifting his self-perception away from centring on the ego and into thinking of himself as part of a network: 'You realise the real question is not "How do you relate to the world around you?" but "How does the world relate to you?"' Sensory triggers lift him out of a state of constant anxiety, reminding him of an experience of Gaza before it was dictated by machines: an illusion of normalcy that only returns with the taste of coffee and shisha, or the scent of a sycamore tree on the breeze. When he's not scanning the city from his apartment window, he's tuning into the news via radio, television, or mobile phone, an activity that absorbs most of the city's residents, setting the agenda for nightly debates that take place between friends and strangers.

Beyond the bloodshed and Saif's own ledger of corpses — those unidentified, dismembered 'body parts of different people mixed up together, lying with flesh they shouldn't be with' in the mass grave of his writing, and those whose names he has learned from those very same debates and broadcasts — life becomes a series of touchstones which coax the mind into trusting every moment of fleeting safety. An orchard's perfume; the landmarks on each block that segment a harrowing walk into mentally manageable distances; the whim to put off shaving until the war ends, certain that he'll be rid of his beard tomorrow, when the strikes should stop. Magical thinking sustains Saif's grip on reality, bunkering his mind away from the unceasing war. 'Boredom has become desirable,' he writes.

04

Falcon

The hierarchy between periphery and focus has been honed over the centuries, shaped by humanity's predatory relationship to the world. Think of seeing through the crosshair, or the viewfinder, compared with a herbivore's atmospheric awareness. 'The lamb learns the world by understanding the world as a system, in all its variation and relation, so that it may effectively remain alive inside it,' writes Anne Boyer, defending the peripheral in her anti-fascist prose poem and extended metaphor, 'When the Lambs Rise up Against the Bird of Prey' (2014). In contrast, the bird of prey 'mistakes its expertise in corpses as proof of its own general acuity'.

* * *

In a sodium-lit hall thick with smoke, a drone flies low over a woman in gymnastic backbend, her breasts bound flat by bandages. The machine's twin green lights appear like animal tapetum in the dark as it cruises below someone walking a tightrope while holding two fans. A girl is held up by a squadron of wan actors, her arms draped around their shoulders like Christ's dead weight; a boy trust-falls from one comrade's back and into the arms of the others. Over the span of 4 hours, the performers let their hair down from spiral staircases, shotgun vape smoke between near-kisses, pour Pepsi-Cola down the walls, spritz noxious spumes of spray paint, and head-bang slowly in a circle, leaping to slide on their knees like soundless rock stars. Perched on a white leather glove — worn on the raised fist of the artist and model Eliza Douglas — a sedate falcon sees it all. Anne Imhof is the director of this opera, the second in her three-part series of durational performances, *Angst I-III* (2016). Equal parts languid, ceremonial, and violent, Imhof's performances have tended towards the immense over recent years, filling enormous buildings — from the Venice Biennale's German Pavilion to the Tate Modern's subterranean oil tanks — with swarms of actors and symbolic objects. They channel sinister yet adolescent moods: the many obstacles between yearning and intimacy; the many violences of watching and being watched. All eye contact is glancing, all movement is hard-edged, no matter if actors are striking a punching bag or swept up together mid-waltz. Imhof's work is synonymous with a kind of brittle, reflective cool. Her coterie of friends and collaborators make up an impenetrable in-crowd — including Douglas, who is also her partner, the composer Billy John Bultheel, and the dancer Josh Johnson. Their allure has brought Imhof into spotlights beyond the art world: magazine covers and editorial shoots. Taking place at Berlin's Hamburger Bahnhof, *Angst II* drew after-hours masses who waited buzzing round the station's entrance, vultures at a nightclub. Once inside they were set loose in its vastness. There's no stage for this opera, nor acts, nor intermission, only Imhof sending directions to 15 actors via text message, leaving her audience to cruise for action. The smallest movement stirs a paparazzi rush. Basal instinct arises, as viewers — even as they raise their phones to record the action, or discuss the performance with their friends — stay tuned to the slightest change in atmosphere.

Imhof has named her characters, a detail that only becomes evident when reading the exhibition text. The falcon is the Prophet, who observes the Diver, the Lover, and the Spitter — the latter a boy whose only task is to chew and spit sunflower seeds despondently. His character was inspired by Violette Leduc's novel *Thérèse and Isabelle* (1966), Imhof says. Set at an all-girls' boarding school, the story describes two rivals: one is preternaturally gifted, the other, as Imhof puts it, is incandescent with self-hatred, and can only 'work really hard... and spit'. The novel's constricted eroticism is folded into *Angst II's* system of symbols and scenes. Sunflower seed husks are mashed into saliva, suspended in a viscous drip. Petulance and indolence play out under intelligent, ever-watching eyes both animal and artificial.

The performance catalogues nihilistic pleasures in an atmosphere of constant ending, channelling the death-wish of late capitalism, its trend cycles and built-in obsolescence. 'Now, as there have been many actions, arts, and sciences, their ends were also many; the end of economics was wealth, that of bomb-making a bomb, that of plastic manufacturing a plastic bottle,' writes Joanna Demers in *Drone and Apocalypse* (2015), describing how every object contains the necessity of its own extinction. New rules have to be learned, new attitudes adopted, but only if you can flip your hair and declare what came before you *dead*. Clad in skull-patterned vests and death metal tees, Imhof's performers enact the most clichéd of glamour paradoxes: the end of youth is youth itself. As doom-laden guitar reverberates throughout the room, the actors adopt a studied aloofness. The attitude feels as balanced on the precipice of expiry as the street-fresh looks they're dressed in.

Someone thumbs a cracked iPhone that will be functionally useless in just a few years, waiting for their next cue. Gillette is lathered across jowls, Adidas swished around contorted bodies. Appearing in Imhof's work alongside ubiquitous brand names, the drone comes into its own as a consumer object. The model that she uses, a DJI Inspire, resembles a dismembered alien skull. Four protruding rotors let it flit insect-like around the room. It looks as cheap and endlessly reproducible as a toy, though it goes for around $3000 new — still a steal compared to the cool $15 million of a Reaper drone. In transferring its core elements from the military to civilian realm, developers have alchemised metal into polymer and death into leisure, yet the commercial drone is still haunted by its origins as a death-machine. 'Drones are still overwhelmingly weapons,' as Adam Rothstein writes in *Drone* (2015). 'Any drone today will look and serve fundamentally as a weapon, regardless of what branding we simultaneously launch.'

Imhof takes this persistent yet diminished violence and adds it to her palette of affects. As it flies, its knife-like rotor blades spinning, the drone's whirr is layered into a soundscape that aestheticises dread. A chorus led by Eliza Douglas accompanies the noise blasted from scattered PA systems and builds to fever pitch. Sometimes a mechanical howl pierces the atmosphere, not unlike the scream of violins that herald a killer's appearance on film, or the low notes of an electric guitar are strummed to elicit the screech of feedback. Fear becomes environmental. Attention is scattered and scraped raw; atonality unsettles even the most mundane of gestures, revealing the danger in everyday objects. Take Demers' words more literally: a bomb can be contained in a water bottle, suicide in a safety razor, hell in the hovering drone.

* * *

In the *Rings of Saturn* (1995), WG Sebald wrote of his childhood ability to sense the world as a complete entity by watching the flight-paths of swallows; the concerted movement of their feathered bodies gave volume to the unruled air. Today, the expanse of the skies — once synonymous with freedom — has been divvied up into commercial routes, no-fly zones, and national borders cast up from the ground. As Gregoire Chamayou writes in *Drone Theory* (2013), aeronautics slice up the heavens, transforming our worldly understanding 'from the horizontal to the vertical, from the two-dimensional space of the old maps of army staffs to geopolitics based on volumes'. Boundless blue becomes faceted as a cut gem.

Lancing through the cavernous Bahnhof, the trajectories of humans, animals, and machines form a similarly complex geometry. 'I've choreographed gazes and hand movements to create perspective in my pieces. The performers are trained to slice diagonals through space with their eyes. Everything else is composed like an image over time,' says Imhof in a 2016 interview with the curator Hans Ulrich Obrist. '[The animals'] function is to create passages, to function like lines parting the space. Those passages were first widened and then narrowed again.' Beneath their thematic designations, Imhof's characters play themselves. Many are longstanding collaborators, contributing music, choreography, paintings, and sculpture to the all-encompassing scenes. All are chosen for the power of their personae, and their magnetic presence — energies that turn bursts of dance explosive, and electrify gestures as small as cracking a can of Red Bull. In contrast, wild animals appear domesticated. Their inclusion isn't intended to pit humanity against animality; rather, they become pet signs, their roles enriched by their place in cultural imagination. In the older work *Aqua Leo, 1st of at least two* (2013), a living mule signalled the abrupt end of a bloodline; in *For Ever Rage* (2015), real tortoises crawled towards eternity, setting the pace for languid performers. For *Faust* — which won Imhof a Golden Lion, the Venice Biennale's top prize, in 2017 — Doberman Pinschers in wire cages trembled with vicious, carceral energy. In *Angst II*, the falcon brims with the potential violence of a predator at rest.

The hawk is historically linked with militaristic aggression, the perfect opposite of the peaceful dove. 'Falconry is the sister of war,' observed Russian ornithologist Georgi Petrovich Dementiev in his book *The Gyrfalcon* (1960). Documented as early as 2000 BC, the practice of using birds in warfare was refined across Asia, where 'eighth-century Turkic warriors were thought to become gyrfalcons after they died in combat; Genghis Khan disguised his armies as hawking parties, and fifth-century Chinese falcons carried military messages tied to their tails', as falconry historian Helen Macdonald notes. Enchanted by the practice of commingling the hunter's brawn with the falcon's precision, Holy Roman Emperor Frederick II of Hohenstaufen penned *De arte venandi cum avibus* (The Art of Hunting with Birds) in the 1240s. Blending Arabic and Latin influences, and combining scholarship from warring civilisations, the tightly structured, scholastic book can be seen as analogous to imperialist desires. *The Art of Hunting with Birds* directed its reader in taming the falcon's unknowable energies, with Frederick highlighting the predator's servitude to its master. No longer an exercise of entwined intelligences, falconry became one of training and domination.

Frederick's manual helped arouse European nobles' interest in falconry as a form of achievable expertise. Soon after the falcon's rise as an elite accessory, it landed in popular consciousness as a form of conspicuous consumption. In medieval England, birds were purchased across classes for status more than sport, though there was still a clear hierarchy of access. Thrice the size of males, female falcons were coveted for their prowess and reserved for nobility, leaving the diminutive males for ordinary people. The falcon would not remain arm-candy forever; the invention of the shotgun in 1850 may have reduced the bird's popularity as a hunting tool, but the development of the airplane in 1903 would once again reconnect it to its history of violence.

Macdonald describes an early aircraft engineer's gaze upwards at a falcon's flight. Neck craned, he sighed: 'When we develop a motor strong enough, that will be the shape of an airplane.' Humans have long coveted the freewheeling freedom of birds; the falcon's speed and striking power also linked flight and violence in our imagination, emphasising the predatory advantages of an aerial view. Bombers and fighter jets changed war forever, turning the sky into a theatre of terror and re-situating humans as helpless prey. Today the military drone can soar high enough to observe its target undetected, then flash down close for a quick, clean strike. Engineers are still looking to feathered creatures to fine-tune aerial developments. In 2010, researchers from the Eotvos University in Hungary recommended that drone pilots closely mimic soaring falcons by harnessing the power of thermals, or naturally-occurring columns of rising hot air, for more efficient flight.

'A lesson common to history, and a truism of capitalism: When humans are motivated by power, they compromise survival, while nature, driven only by an impulse towards survival, manages to retain its power,' writes cultural critic Ayesha A. Siddiqi, captioning a photograph of a golden eagle snatching a consumer drone out of the sky. The picture was taken during a 2016 police training exercise in the Netherlands, originally accompanying an article describing how French and Dutch police have drafted the birds as an anti-drone force. Though many models are pre-programmed to avoid restricted airspace, these settings are easily overridden, and consumer drones have so far shut down airports, aided assassination attempts, and been rejigged to carry explosives. The bird of prey programme was later abandoned — the blade-like propellers of larger drone models outmatched the eagle's talons, risking injury to the birds — but the hope that wild instinct would triumph over technological sleight, however short-lived, lit up an escape path for Siddiqi. The drone emblematises a power dynamic that renders the individual as small under networked technology as a field mouse under the hawk's eye. She reflects on simple ways to evade such paralysing complexity: tying a mask over your face to escape surveillance, throwing a rock through a screen to disable it, or slipping the net simply by leaving your phone at home. 'I don't care about power, I care about survival, and because I'm not afraid of dying I care about your survival more than mine,' she continues, echoing Boyer's pro-prey sentiment. 'Which makes me a bad person, but a good animal.'

* * *

In search of signs of life, *Angst II's* viewers roam the imposing hall of the Hamburger Bahnhof, circling the bone-white spiral staircases, brushing past the white leather punching bags, decorated with drawings of enthusiastic fellatio, that dangle from the rafters. In between action the performers bend and stretch or flop together, silently commiserating or checking their phones. ('Hold a pose until you are bored with it,' Imhof instructs them, pre-performance.) Bursts of activity seem spontaneous, as sudden as birdsong: an arm wielding a baseball bat is wrapped around another neck; someone is shaved by a double-edged razor; the tightrope walker once again walks the line pulled tight across the room.

While she was studying art at the Städelschule in Frankfurt, Imhof worked at the nightclub Robert Johnson, where the candy-pastel interior and friends-mostly policy left a lasting impression. She has spoken of her love for blurry saturation, as with encroaching dusk; for atmospheric intensities and 'ghosts that become your friends and accomplices', in true club-kid style. Who, in the nighttime fade, doesn't feel all soft surface? What's a club but a shallow watering-hole, or a collection of vapours in an empty room? Imhof sees what happens when the entropy of hedonism is paused, isolating its elements and stretching them out. Some of her performers are models hired from a Berlin agency; at intervals, they break into catwalk sashay and moshpit contact. In the nightclub, on the runway, and under the drone's eye, it's all occasion for display. Think of what Baudrillard incanted in *Seduction* (1990): 'seduction never belongs to the order of nature, but that of artifice — never to the order of energy, but that of signs and rituals.'

Since Imhof's youthful heyday at Robert Johnson, synthetic love has been swapped for hyper-performativity, complete surrender for a climbing, tangible sense of popularity. The furtive watching that once held the possibility of connection — eye contact across a darkened room — is now unflinching yet stunted by the screens that facilitate it. The substances scattered throughout *Angst II* supplement the anxiety that tails contemporary hedonism, Red Bull and Pepsi Max dissolving hangovers and ratcheting up heart-rates, plumes of vape smoke calming nerves scattered by constant exposure. It's easy to imagine how speed might persist in this universe, stretching out stamina rather than producing any euphoric feeling. (And how, in this age of acceleration, could we resist something so aptly named?) Molecularly close to Adderall, the prescription pill used to achieve the same ends at work and, off-label, at play, it makes up a regimen of the hyper-productive made vacuous by burn-out. Seen through synthetic highs, violence and pleasure aren't so much opposites as bedfellows — a relationship that Imhof underlines by choreographing jabs and feints with a spiky eroticism. Nazi soldiers gobbled amphetamines to 'banish the fear of combat and replace it with a sense of invincibility', as drug documentarian Mike Jay wrote in his essay, 'Don't Fight Sober'. A speed pill called Captagon, embossed with rough yin-yangs or double Chanel Cs, made headlines in 2015 as the fuel for 'zombie' soldiers roaming Syria, 'all smiles, across fields of ruins and severed heads'. Military drone pilots maintain their focus with energy drinks and Adderall. The chemical cocktails sustaining their bodies are as synthetic as the polymers that make up the consumer drone. In *Angst II,* this continuum is made obvious: fizzed syrup, hovering machines, and slack bodies all become proximate.

It's possible to see *Angst II* as one giant bedroom, if one thinks of the bedroom now as retreat, cage, and asylum. Like teenagers or inpatients marking the hours with idle scrawl, actors drag a single line of paint from one wall to another, or scribble on the white ground in gold marker, or concoct dark paintings with scratchy handwriting and washes of black. Cans of soda are popped and hardly touched. The majority of the performers are thin, white, mussed from the bedsheet's solo toss; their aesthetic a snapshot of the proclivities of the moment. The fashion Imhof channels is a brief history of techno-culture, from t-shirts bearing shiny CG waveforms to the nostalgic roll-necks and wire-framed spectacles of Bill Gates and Steve Jobs. As the West slowly awakes to its self-created dystopia, these performers are dressed for the occasion.

Critics have described *Angst* as a 'sad teen' performance for the ages; the choreographed gestures, stretched to their limit on the rack of time, recalling adolescent slouch and squalid depression. Boredom, my friends all say, is 'early death'. Too much of it brings madness, or the existential fear that follows vertiginous awareness, but embracing it is also freeing — the dead, after all, bear no responsibility. If 'the whole earth has turned into a malicious, hopelessly divided middle class that is doomed to perish', as Dietrich Diedrichsen writes, considering the ruins of counterculture in *The Whole Earth: California and the Disappearance of the Outside* (2013), it has happened even as the middle classes' traditional rewards — job security, home ownership — have evaporated. Lacking any real imperative to grow up, yet still under pressure to produce under the old world's rules, it's no wonder that youth is intensely fetishised by adults who aspire toward looking good and doing nothing. Leisure has become the only escape from a hostile yet protected world. 'The urgent sense that we are doing something, that *something is happening*, temporarily overrides the idleness at its core,' writes the critic Alison Hugill of the performance. 'The labour generated in this piece represents a means without end, which — in a capitalist society where value-producing, waged work dominates — might be the most unnerving kind.' In the dirge of witch-house band Austra: 'In mornings I rest here / In evenings I work here / My debt isn't spent.'

Above the action at *Angst II*, the drone lingers on the slack loop of its own flight, hovering like a wallflower. An unwelcome guest, it's entertained nonetheless. Writing about the drone's presence, Éric Troncy considers 'the performance as a warzone', and one can see the link to the military drone's persistent leer, waiting for a quarry's telltale gestures to unwittingly signal a strike. But the connection is a reach. Far less is at stake, and it's all too murky for direct analogy. Instead, Imhof has described how human and nonhuman actors perform 'like layers, on top of one another, like applying oil paint on a surface'. Fear can be hot, as long as the body's protected. Privilege invites recklessness. See the exhibition from a satellite view — zoom out of the hall, out of the city, out of the country — to the performance's dead-cold register on the scale of relative tensions. Nihilism is easy when there's nothing at risk, and beyond its treacle-drip seduction, *Angst II* is a monument to what's pursued under safety — the suburban delinquency and aestheticised death-wish of every overgrown teen.

05

Field

On the outskirts of Malaga, a cyclist swoops through hairpin bends, his white jersey showing bright against the ochre scrubland as he's filmed from above. Edited to a crisp pop beat, the minute-long video is for an independent cycling team — one of many cataloguing their races and downtime, mainly for the viewing pleasure of other avid athletes. The drone operator is my younger brother, who cycles for them too, though here he has driven ahead to send his toy-sized model into the air, flying it back along the serpentine track. I know that the alternate shots made with a handheld camera — floating alongside the cyclist's quads, which churn like a couple of leaping dolphins, or looking him straight in the gasping, soundless face — were filmed riding backwards on a rented tourist scooter, and that the team fuelled up mid-ride with plasticky Swiss rolls bought at gas station cafes. It was an ad-hoc shoot, in other words, but the end product was sleek and swift in the style of contemporary advertising, full of pulse-quickening turns and jerseys shrugged on in golden-hour slow motion. The film's secondary subject was asphalt, laid out in mouthwatering, car-commercial curves, spun from an invisible axis as if the landscape were not anchored to a planetary mass, all from an aerial view.

The drone views serve a purpose, counteracting the high-octane grit of rider-mounted footage and allowing vicarious onlookers to orient shaky, bipedal action within a more elegant and cohesive world. For now, my brother flies the drone himself, riding shotgun in the team van, but this particular skill could soon be rendered obsolete. As of this book's writing, there is an emergent genre of self-filmed sports videos, enabled by new consumer drone models made by brands such as Staaker,

Skydio, or DJI. Whether they're tracking skiers down slopes, surfers cutting up waves, or cars drifting on a dirt track, these drones are linked to their athlete via a wearable sensor, or use rudimentary artificial intelligence to pick them out of their surroundings. Given the drone's connotations, I feel a prick of anxiety when thinking of wilfully teaching a Skydio who I am so that it may better follow me — even if it is harmless in practice, accessible enough to the serious-minded amateur at $999, operable via smartphone, and thrown into the air as easily as a hawk from a falconer's glove. Directing its hunt by registering human bodies in its sensory system, the self-flying drone brings the observational terms of the battlefield to the playing field, bringing us back to the centuries-old affiliation of sport and war.

After the Drone Papers were leaked by the online news publication The Intercept in 2015, detailing the US military's covert assassination programme in Afghanistan, Yemen, and Somalia, the US press became momentarily fascinated by the codewords used to deal out the killing. Some of these expressions had to do with gambling, others with sport. 'Touchdowns' are kills, for example, as are 'jackpots'. The difference between the two is the nature of the targeting. Many victims are tracked via their mobile phone signals, and 'touchdown' means that the signal itself has been snuffed out. 'Jackpot' is less ambiguous: it defines a confirmed kill. To line up the hits, higher-ups such as former president Barack Obama consulted 'baseball cards': one-page visualisations of a target's pattern of life. These printouts were designed to be read at a glance, completed by a headshot and an aerial view of the victim's place of residence. These metaphors, however mixed, aren't wholly incompatible with the contemporary battlefield. Sport has long been connected to war, an interdependency that's particularly conspicuous in the United States, where football coaches give speeches cribbed from Black Hawk Down (2001), Ridley Scott's film about the Battle of Mogadishu, and even golf's stately pace is given a hawkish turn, such as when fighter pilot Major Dan Rooney was enlisted to prepare the US team for the 2010 Ryder Cup. 'Army and Navy endorsed the old General Wellington idea that battles were won and lost on the playing fields of youth,' wrote Steven Stark in an article exploring this very theme for *The Atlantic*. 'The better the sports program, they reasoned, the better the soldier, in a reverse of all those recent team talks.'

In 'Light of God', which appears in *Acid West* (2018), a collection of autobiographical essays on the derangements of the American Southwest, Joshua Wheeler spends the length of a baseball game thinking about the lives of drone pilots, who sit in air-conditioned shipping containers just a short drive from the stadium. Paunch bellied, heads monastically bowed under 'text neck' — the particular pain that results from looking too frequently at one's devices — these Average Joes dub their home state 'New Mexistan', not least because Hollywood has already co-opted the local landscape for the production of combat blockbusters. The desert becomes the blank screen onto which the US's perpetual wars are projected. It is the backdrop for drone pilots' day-to-day life: it's also the stage set for multi-million-dollar films, such as *Lone Survivor* (2013), that glorify violent US interventions. Studios painstakingly recreate the architecture of Iraq and Afghanistan, the very places that the pilots are monitoring and bombing remotely just a few miles away. From his seat in the grandstand, Wheeler sees how the belief in good triumphing over evil — of being on the winning team — is deliberately woven into white American identity, stoking the narrative of an 'us' versus 'them'.

'The awful great thing about baseball is that it's boring to watch,' writes Wheeler. 'I can get lost in pondering all of existence, but I've got the crack of the bat to snap me back into the story...before I drift again into a lazy anxiety about the universe.' Such intentional idleness, we gather, is also the experience of the drone pilot; the 'crack of the bat' is equivalent to the lightning-strike of Hellfire missiles, the seconds of action that validate the doldrums of days, weeks, or months. Though he makes drone pilots out to be as dumb and reactive as sports spectators, propping up their attention spans with fistfuls of Doritos and M&Ms, Wheeler doesn't truly criticise the Americana that he so enjoyably participates in. He thinks fondly back to when Nighthawk fighter jets used to fly in formation over the baseball diamond on special occasions to rounds of applause. 'I don't guess we will ever get a triumphant drone flyover to commence a baseball game,' he laments, mourning lost valour and patriotism rather than lives. He always speaks of the drones as 'ours'.

As with Wheeler's tale, it has become commonplace for pundits to compare killing from a distance to ball- and videogames. Combat videogames, in particular, establish a deliberately casual relationship to death, scrubbed as they are of even the one-sided tear-jerking of Hollywood propaganda. Killing others is the goal, and if you die, you're shuttled quickly back into the world of your origin, your finely rendered gore-splatter vanishing without a trace. Over the years, the drone wars have been publicly processed as a 'costless and bloodless exercise', as Eyal Press writes in the *New York Times* in 2018, perceived to be the work of '"joystick warriors" engaged in an activity as carefree and impersonal as a video game'. He points to how Philip Alston, the former United Nations special rapporteur on extrajudicial executions, 'warned in 2010 that remotely piloted aircraft could create a "PlayStation mentality to killing" that shears war of its moral gravity'. Yet the mentality is not limited to drone pilots alone, and the implicit criticism of the gaming industry — rather than, say, the historic relationship between war and masculinised leisure — may prove a red herring. An 80-year-long study conducted at Harvard University from 1939, which followed the lives of

268 men in an attempt to uncover some truth about happiness, inadvertently documented attitudes to war. 'The whole thing was like a game,' said one subject, referred to as Case No. 141, of his time in the Air Force.

* * *

A promotional spot for the Staaker self-flying drone details the various automated compositional modes available to a videographer: 'hover', where the drone remains stationary though its camera follows its target, an eyeball rolling inside an unmoving head; 'follow', which throws it into hot pursuit of its subject; 'compass', a side-on shot that stabilises what's happening in the lower third of the frame, allowing the landscape to fill the remaining two-thirds; and 'circle', which sees it revolve around the athlete, lending them an air of 360 majesty. This strain of vision has an analogue in third-person console games — think *Tomb Raider* (1996-2018) or *Grand Theft Auto* (1997-2013) — where the illusion of an invisible camera tails a player's character so that they can see their whole body as they run and tussle through a fantasy landscape. The 'camera' finely recalibrates its movements to maintain the correct perspective, placing the subject, inescapably and quite literally, at the centre of their own world. (Such an approach differs from first-person games, where the 'camera' is located inside your character's head, your two arms visibly extended before you. In these, vision is closely linked with action; you move in the direction you look.) Fundamentally dissimilar to how we experience reality, the third-person view is closer to an out-of-body experience, the kind of dream in which you are in control of your body yet floating somewhere behind your own head, watching your life play out from a mysterious remove. In her short video works, the artist Anna Mikkola exploits everyday machines' usefulness with tasks or environments that strain their sensible, built-in logic. Having first trained as a designer at the Bergen Academy of Art and Design, Mikkola is as well-versed in how form begets function as she is in how susceptible design-led thinking is to subversion. Her videos feature among a rich oeuvre of installations, sound pieces, and performative lectures, which tend to analyse systems both large and small. In *From Simple Parts to Collective Complexity* (2018-2019), a real-time weather simulation is triggered by the movement of live ants travelling between transparent vitrines, drawing elegant causality between infinitesimal actions and major environmental consequences. Her interest in scale works the other way, too. The 5-minute video *Blind Spots* (2017) sees an automated domestic vacuum cleaner set loose in a half-constructed parking lot, a terrain specifically chosen in order to vex the elaborate system of sensors, feedback, and data collection that enable the machine to perform its task. Textured concrete, on-ramps, and dirt boggle its singular 'desire' to traverse and clean a room. A camera mounted atop its body captures the derangement of its sensors: the jagged, spinning movements as it fails to fulfil its directives yet never receives the cue to stop.

The later video *From A Distance* (2018) draws similar poetry from automated vision, this time using the self-flying drone's determination to tail whomever it recognises as a subject. Its construction is relatively uncomplicated, filmed entirely with a drone, which tails a person running in circles around a clearing. A thicket of trees hems the field on all sides, their leaves turning orange in the fall. In the half-light, they become a dense and uncrossable boundary, closing the park in from the city, a place we glimpse occasionally as bursts of white-washed houses and winking lights as the drone floats up and down. A single action taking place within a fixed area, the video resembles Bruce Nauman's studio exercises, such as *Bouncing in the Corner* (1968) and *Walking in an Exaggerated Manner around the Perimeter of a Square* (1967-68). Convinced that any activity occurring within the walls of his studio was art, Nauman recorded himself engaged in various nothings, repeating a particular gesture for an hour — the length of an entire videotape. For *Bouncing in the Corner*, he sprung up and down on the balls of his feet for the tape's duration; in *Walking in an Exaggerated Manner*, he lunged and recoiled from corner to corner of an invisible square, marking the shape out with his movements. Only in Mikkola's work, the drone unsettles the prosaic infinity of jogging in a circle. The runner is trapped in a tightly-bordered world from which there is apparently no exit. As sunrise soaks the perimeter, bleeding in incrementally, the machine cuts after the running figure, insistently centring them in its visual plane, self-correcting as it flies.

To create *From A Distance*, the drone first had to map the key elements of its subject — in this case, an anonymous figure dressed in black, hood up — and then determine the correct height at which to best fix them in its sights. It's hard not to see the video as the document of a pursuit, especially given the black hoodie's potent symbolism and the drone's need to define a target according to a set of oversimplified characteristics. In the US, the hoodie became a symbol of racial injustice after the death of Trayvon Martin, who was 17 years old, unarmed, and wearing the garment when he was fatally shot for looking 'suspicious' in 2012. Mikkola milks the sense of unease with heavy breathing — her own, recorded during an evening run — superimposed over the runner's orbit. A lag, as small as a keyhole incision, opens up between two realities: as the runner bends to lace a shoe, the sound of a jogger's breathing continues, as do the footfalls and regular swish of technical fabric. Until this moment, it's easy to believe that runner and breather are one and the same. Now, the video makes us wonder, who else is out there? Who is stalking the fugitive runner? It serves as a reminder of all kinds of watchers, from policemen to drone operators to the third-party services tracking our actions, mingling with our own thoughts and bodies.

* * *

When I spoke to Mikkola on the terrace outside her Somerset House studio, evening was fast approaching. The wind had grown cold, distributing the mist from the fountains onto our hair and faces, and littering small particulates whisked up from the surfaces of the city on the foam of our drinks. Mikkola had been preparing the lecture component of an upcoming performance in the Netherlands, and our conversation roamed around its central themes: automation and intelligence. We discussed the making of *Mimicry in Formation* (2018), a virtual-reality experience that formed one element of the installation *Life is Necessarily Complex* (2018). To access it, a viewer straps on a headset that dangles from a length of thick black chain. Once 'inside', they find themselves transformed into a migrating bird, wings churning in the centre of its flock. The sky is darkly rendered. Clouds streak by in luminous, multi-toned whites. Land and sky stretch, edgeless, all around. Exploring their field of vision, the viewer experiences the vertiginous feeling of looking down to see no feet planted firmly on the ground. And then the plunge begins, vision spinning out, the landscape torrenting upwards and through their shattering bird-body. A glitch breaks the mountainside into fatal spines which spear the undying viewer. Never to experience a realistic impact, they continue to fall straight through the earth.

The fall, Mikkola said, wasn't intentionally coded. Instead, a stray line wove it into her tenuous new reality, and she decided to keep it, impressionistically toying with the medium as a painter who surrenders to accidental swipes or drips.

On the terrace, she describes the influence of third-person games such as *Assassin's Creed* (2007-2018), which take place in 'open worlds', allowing players to roam fantasy landscapes freely rather than forcing them to adhere to linear, goal-oriented gameplay. Boundlessness, however, is an illusion — the 'camera' stubbornly revolves around your character, blocking your full view of a world rendered specifically for your pleasure, and 'invisible walls' fence off unplayable areas. The games granted Mikkola piercing moments of beauty that seemed to be heightened, rather than dulled, by these restrictions. They piqued her interest in how technological flaws might unearth some unexpected poiesis, inspiring *Mimicry in Formation* and *From a Distance*: how a virtual-reality glitch might cast a new, surreal spell over you; how the functional limitations of a videogame might offer commentary on personal autonomy.

In name, 'open world' games sound like they should offer limitless possibility, yet many channel the same old themes of war and empire, with even the most innocent premises dependent on discovery and acquisition. Though the first of these were developed in the 1980s, the genre was popularised with the release of *Grand Theft Auto III* (2001), which continued in the 'guns, gangs, girls' tradition of its prequels, leaving players free to terrorise a fictionalised New York City. The later *Grand Theft Auto V* (2013) was set in a facsimile of Los Angeles and heralded as containing more hedonistic opportunity than ever before. Its creators were inspired by how 'movie stars, migrant workers, and criminals have an equal stake in [the city's] cosmology', contributing to its aura of individual triumph both illustrious and illicit, as Sam Sweet wrote in the *New Yorker* upon the game's release. There was one crucial difference, however: 'Los Santos', as the fictional city was called, was made into an island 'to dispel any notion of unreachable territory', Sweet writes.

Intrigued by the game's metafictional nature, the experimental filmmakers Larry Achiampong and David Blandy moved their *Finding Fanon* series (2015-2017) from the real world into *Grand Theft Auto 5*'s virtual territory. The films take cues from the writings of anticolonial theorist Frantz Fanon, who famously saw violence as crucial to decolonisation, but their content is meandering rather than militant. In *Finding Fanon 2*, the first episode of the series to be set in *Grand Theft Auto 5*, avatars representing Achiampong and Blandy travel to the game's farthest-flung scrubland and shipping ports. These areas are eerily still, empty of the globalised bustle that sustains real-life Los Angeles. Shipping containers are stacked untouched, and no 18-wheelers grind down the wide roads. The characters dive into a rain-pocked ocean and swim towards the horizon until they come into contact with the game's invisible wall — the point at which these seemingly infinite worlds come to an abrupt end, the rest of the landscape continuing inaccessibly to the horizon. An 'open world', they discover, is a paradox. Much like a nation that only offers personal freedom to those entitled to participate in its nationhood, the game's plenitude is defined by a strict zone of exclusion.

Mikkola probes the nature of these expanded yet narrow universes, taking moments of friction, which undermine otherwise seamless experiences, as conceptual directives. In *From A Distance*, she transposes the relationship between player, boundary, and camera into real life, using the natural boundary of the woodland as her scenario's 'invisible wall'. Caged by the open field, the runner could run around it forever; the drone, programmed to follow from the optimal distance to capture both figure and landscape, could continue to pursue without the conclusion of a strike or crash. The patched-in breath could continue to circulate, its cadence cyclical and calm. But this set-up, we eventually come to realise, will not follow the hermetic logic of gameplay. When the runner drops into a low crouch, the drone's sensors are confused. It rises into the sky to recalibrate its perspective, taking our view with it. In this instance, our sense of passivity is complicated: was it our gaze that had trapped the runner all along? Out of sight, they can give the drone the slip, breaking the spell of inertia.

06

Spirit

You said imagine the near future
when our eyes have moved to the sky
we will be looking down at each other
a connectivity equal to an escape.
Korakrit Arunanondchai, Painting with History in a Room
Full of People with Funny Names 3 (2015)

The cab's interior was pink and red. Over its glossy white paint job, a decal read 'ILLEST' in blood-drip horror paint. I entered the car, which I had ordered to my location on my phone, with the sense that I was climbing into an open mouth. The gold ballcap on the back dash resembled a single dental crown, my body perched on the humid leather like a pill on a tongue. We ascended one of the city's myriad flyovers, glassy buildings and dark vegetation giving way, momentarily, to a gilded temple rooftop as MIA bounced on the stereo. I had landed in Bangkok at the end of monsoon season, the air thick with incumbent rain, with the intention of seeing Korakrit Arunanondchai's *Ghost* festival of moving image and performance. *Ghost* was curated around its eponymous apparition, encompassing installations by 12 artists — among them Apichatpong Weerasethakul, Hito Steyerl, WangShui, and Ian Cheng — at nine galleries across the city, along with a programme of screenings, parties, and performances, the last of which would take place around a thousand year-old sacred tree.

Arunanondchai himself would not be exhibiting, but I wanted to see the world that his work lived in: the atmospheres that informed it, the people that passed into and out of it. I was intrigued by the way he pursued emotion itself as a discipline and a form, spinning friendships, folk mythology, technology, geopolitics, and poetry into total works of empathy. Each one of his videos, performances, or installations convinced me anew of our togetherness, lighting up the fine threads that yoke visible to invisible worlds like a torch shone on a spider's web; he returned storytelling to its collective nature, showing how we never arrive at ideas alone.

'Ghosts could be phantoms, feelings, or invisible systems,' his curatorial letter for the festival read. I felt it. Later, WangShui will tell me of the week spent installing their video, *You Belong Two Me* (2018), at the Jim Thompson House, where electrical equipment mysteriously malfunctioned and the staff all dreamed the same unsettling dream, until an offering was made to the house's resident spirit. The apartment I had rented was halfway up a high-rise, air-conditioned to subzero temperatures, and new enough to be free, I hoped, of paranormal activity. Every morning, I bought a can of coffee from the 7-11 below my building and shot the black syrup in the backseat of a taxi as I hurtled towards another unknown. Today was no different. Inside the cab, I scrolled past a tweet generated by one of a number of esoteric poetry bots. Other such bots wrote botanical sexts, or picked the day's gender from an infinite array of affects, or determined an ambiguous 'threat level'. This one wrote small, second-person dramas set in deep space. 'You consult your maps against the stars set before you. You pick a point of blinding light to the South and engage your plasma thruster,' it read. The year was 2561, or 2018 CE. Outside a low-slung complex housing a number of galleries, the scent of charcoal-grilled meat wafted across the crowd. Everyone sat attentively on plastic stools, clutching gnawed BBQ skewers, as a triple screening of films wrapped up. Curator Aily Nash had selected the films — *The Human Surge* by Eduardo Williams, *Kempinski* by Neil Beloufa, and *Mundane History* by Anocha Suwichakornpong — for their shared disembodied perspective, accentuating the eerie view with the screening's title: 'Spectral

Vision'. As I played catch-up, someone offered me a cup of beer on ice. The artist Stephanie Comilang wandered through in a pair of plastic carp slippers — an ingenious foot-in-gill design, worn so that one's stride resembles the creature's flip-flop out of water — as a film scholar named R flagged me down, recognising me from the emails we'd exchanged shortly before I arrived. She introduced me to A, a researcher of 'forest communism'. The forest or jungle has been a gift to revolutionaries, guerrillas, and other minority groups who wish to maintain their independence, they said: revolutions have been won in collusion with that opaque biome. Some scholars have attempted to define the dense woodland shared by Vietnam, Laos, Myanmar, Thailand, and southwest China as a discrete cultural region, unified only by its hostility to unity. Zomia, as it's called, deviates from governance. It obstructs projects of nationhood, distorting party lines. Offering safety, invisibility, and bounty, yet heaving with unpredictable danger, it's little wonder that the region has spawned countless myths, superstitions, and feature films. By no coincidence, its power is an undercurrent that runs throughout Arunanondchai's work.

The jungle still half-reigns over the city, its twin and antithesis.[2] Plants proliferate in any space neglected by the unceasing construction and demolition of buildings. Against the sheer force of biological life, concrete seems as impermanent as a track in the mud. Residential gardens become quickly overgrown. Public parks retain a sense of untameable lushness. The tropical metropolis is well acquainted with entanglement, famously magnetic to those who prefer chaos over control. Arunanondchai — who moonlights as a rapper, his occasional performances charged with coarse, frenetic energy — wrote a song about its unique character. 'Bangkok CityCity' opens his video *Painting with History in a Room Full of People with Funny Names 3* (2015), his vocals pitched-down and hypnotic. After a montage of sunsets, chicken-munching crocodiles, and jet-skis cutting down the Chao Phraya river, a drone pulls up and over a denim-clad gang. Alternately posed in the backs of pick-up trucks parked in gridlock, or vamping on a skyscraper-topping helipad, they mouth his lyrics — an ode to a city that's 'a fantasy'.

The *Painting with History* series is centred around a character named the Denim Painter, played by Arunanondchai. 'He is basically just me, but more like the art character of me,' says Arunanondchai in conversation with the curator, Robin Peckham, for *Leap* magazine in 2015. Peckham observes that, to an extent, all public-facing practices require a designed persona; the Denim Painter can therefore be read as both a fictional character and a metafictional manifestation of Arunanondchai himself. The videos' cast of characters includes the performance artist boychild, appearing as herself, and a drone-spirit named Chantri, voiced by Arunanondchai's mother. As he embarks on various projects of discovery and actualisation, he relates his experiences to Chantri in letters that make up the videos' subtitled narration. The epistolary form is intimate, well-trodden ground for the essay film — think Chris Marker's *Sans Soleil* (1983), with its disembodied narrator's wistful, reflective tone — and it lends itself well to Arunanondchai's inclination to process world and life events together, allowing the videos' narratives to flow down associative, nonlinear paths. Even when she is right beside him, or hovering just out of reach of his outstretched hands, it feels like they're addressing each other fondly from a distance. He speaks Thai; Chantri replies in French. Their relationship is defined by its conduct across multiple, blended worlds — East and West, human and nonhuman, physical and spiritual.

When we first encounter him in *Painting with History 1* (2013), the Denim Painter is dressed in jeans and matching jacket, both splattered in red, yellow, and blue paint. He favours denim because of its subtext: workwear elevated to fashion, globally disseminated yet emblematic of big sky America thanks to decades of advertising and branding. Beginning his journey in the US, where he attends a liberal arts college before returning to his home city of Bangkok, he toys with how best to portray his ideal persona, selecting and shedding potent symbols whose meanings fluctuate between the false duality of East and West. At first blush, he seems given to the wilful naivete reserved for boy artists, unembarrassed to speculate about the nature of art and the universe. His desire for recognition is pathological; he cuts himself into a deluge of recent and historical footage with the partial aim of putting his work into context. Clips of Pollock, Richter, Warhol, and Kiefer mingle with those of monks scorching primed surfaces with open flames. Considerable airtime is also dedicated to Duangjai Jansaunoi — the pop culture sensation and *Thailand's Got Talent* contestant, who rose to divisive national fame by applying acrylic paint to her naked torso, then her torso to the canvas. To the Denim Painter, these disparate influences share an expressionistic vision that he channels in his own work. Bleached and scorched denim is stretched canvas-taut, bearing hand-and body-prints that replicate Jansaunoi's application. These are shown and sold by Arunanondchai, once again blending real and fictional artistic personae. I understood some of the paintings' elements to be bad-on-purpose, appropriating Jansaunoi's bathetic body outlines to highlight the Denim Painter's solipsism — or else to mock an attachment to modernist masters, conflating the canon with reality TV stardom — then sold off as props that retained the aura of their screen time.

Where another artist might see their tools' commercial context as purely incidental, Arunanondchai is as invested in the drone's ordinariness as he is in that of blue jeans or acrylic paint. He chose Chantri's model, the DJI Inspire, exactly because of its 'prosumer' standing, beloved by amateurs for both ease of use and image quality. If Chantri suffers a collision, or drops out of the sky into the ocean, no matter; her body is fully insured, so it can be entirely replaced. Just as he highlights denim's significance as a global uniform of casual belonging, he sees the DJI drone as a conduit of communal vision. 'If it's a drone that everybody uses, you can seamlessly use other people's experiences as well,' he tells me, much later, when we finally meet. In other words, you can share their minds. Throughout his videos, Arunanondchai splices in island and urban panoramas taken by other enthusiasts. These are the visions by which Thailand broadcasts its fantasy the world over. The *Painting with History* series draws oblique parallels between the Denim Painter's search for meaning, authenticity, and self-actualisation as an artist, and the similar quests undertaken by tourists, devout worshippers, sports fans, and political activists. 'Connectivity', as he says, 'is equal to escape.'

'The you of the future may collect the us in the present,' says the Denim Painter to Chantri when he watches a sunset through her eyes in *Painting with History 3*, giving away his longing for belonging. He speaks in a succession of aphorisms that form the video's narration, musings on cosmic themes and national politics alike.

* * *

'Where is Krit, anyway?' I yelled over a mismatched club mix of Britney Spears' 'Toxic' the following night, using the pet name that everyone seemed to call Arunanondchai — a testament, if anything, to his expansive social circle. I was irresponsibly leaving most things up to fate, tagging along with A and R or befriending other *Ghost* attendees when we shared taxis between venues, and had banked on catching Arunanondchai at the party. 'Don't worry,' R said, handing me her drink; in a few seconds, she confirmed that he would meet me the next morning, her fingers moving lightly across her phone screen. In the meantime, it was 1am and Maggie Choo's — a popular drag bar in what was once a bank vault, now the basement of an unassuming Novotel — was closing. The performers had concluded their vamp down the makeshift aisle, scarlet fans flashing open between taloned nails like semaphores, and the music was taking on the dance-hostile tempo that meant it was threatening to cut. So we followed A outside, first to a tourist-trap crawling with stray cats, and then to a bar in the embassy district rumoured to be the drinking hole of American spies.

If the gossip was true, the spies certainly flaunted indiscretion. Polaroids of bar guests, always with one arm slung over the owner's shoulders, plastered every inch of the narrow, smoky room. The rest of the decor was equally deranged, from the sliced pickles scattered loose in the urinals to the TV looping softcore pornos, and the lighters thumbtacked to the wall above each table, swinging from bits of plastic twine. We'd barely arrived when a man lurched at us from the depths of his blackout, but I felt sheltered by our burgeoning friendship as A talked circles around him, joking him into submission. Besides, the sprawling evening felt apposite following *Ghost*. All of Arunanondchai's videos ripple with the sustained pleasure of an endless hangout, a sociability that is integral to how they're made. 'During my research phase, I have conversations with people who share my interests or have knowledge in certain areas, whether it's technical, scientific, or spiritual. I might talk to a shaman or a scientist, or my brother who is a VR/AR engineer,' says Arunanondchai to curator Martha Kirszenbaum in *BOMB*. The scripts he writes, using these discussions as source material, also take the form of conversations between the Denim Painter and Chantri, or the Denim Painter and boychild; understanding is always formed in dialogue.

But as much as Arunanondchai's oeuvre speaks to late-night experimentation, there's a counter-flow, too — a sense that while he (or the Denim Painter) relishes being on the scene and enjoys its usual rewards of notoriety, attention, and prestige, part of him is always back home with his family, touchingly aware of their thoughts and wellbeing. The Denim Painter's dialogue is replete with the anxieties that attend unconditional love, as he negotiates how to balance his individuality with his desire to do good by them. In a review of the New York writer Jenny Zhang's short story collection, *Sour Heart*, Emily LaBarge defines the family unit as a place of innate understanding that nonetheless exacts its own demands on a person. 'One is owed and one also owes unfathomable quantities of love – a tie that both buoys and binds,' she writes.

It's a phrase that I returned to, rewatching *Painting with History 1* to 3 on my laptop as I ate a cup of instant noodles harvested earlier that day from the 7-11, the sky lightening with the first inkling of dawn. 'My father called to talk to me about love, but I was too busy,' the Denim Painter says in *Painting with History 1*, an attitude he comes to question by the middle of *Painting with History 3*. 'Chantri, are you recording?' he asks, as the drone hovers bedside at a hospital. Chantri is monitoring Arunanondchai's grandmother in his absence, whose head is nestled into a polka-dot pillow, hair soft as a spent dandelion. He thinks aloud about the fragility of human consciousness — but why isn't he there, at the hospital, himself? Torn between self-annihilating familial love and the pull of his ambition, the Denim Painter gravitates towards the latter and stays haunted by his choice. Arunanondchai himself is clearer about the interrelation between his family and creative lives, saying in *BOMB*:

The reason I started videoing was because I wanted to find a way to hang out with my grandfather. That was in 2011 when he was losing his short-term memory and I was trying to keep a close relationship with him. I started to film him, and then I would show him the footage after the edit. Every time I visited, I had a video for him…Basically, my grandfather's Alzheimer's got me into filmmaking.

In an essay for *The New Inquiry*, the cultural critic Maryam Monalisa Garavi writes that the drone's most human element is its 'paranoia'. *Painting with History 3* shifts that paranoid mindset to the drone's human partner, who is unsettled by his own finitude, his inability to be everywhere for everyone at once. In comparison, Chantri is an extended eye and an augmented mind, promising posterity as she records what he already knows he's doomed to forget. 'We build things to keep memories,' he says in the video's prequel, referring both to the prosaic fragility of his smartphone and the nature of memory under Thailand's military junta, which has been notoriously censorious during its 5 years in power. 'But today, it seems harder and harder to keep them.' Like an airborne iPhone, Chantri is an external brain that collects and broadcasts isolated expressions of love. Unlike an iPhone, however, she also provides presence. Fully sentient, even wise, she goes to where the Denim Painter can't physically be, helping him divide himself between two seemingly irreconcilable existences, or floats by his side, a faithful companion instead of a weapon or spy. Infinity, however, is an illusion. Later, she'll fly over towering heaps of computer waste left to rot in the Thai jungle, reminding him, and us, of the inevitable obsolescence of her physical form.

* * *

If you appear in the poetry
you will be located on a sweeping landscape
as if neither they nor myself have ever existed,
or else, we are omnipresent. Such absence
is possible only in imagination, while
omnipresence may exist in virtual reality.
Zhou Zan, Her 她 (2017)

Chantri wasn't always a drone voiced by Arunanondchai's mother. In 2013, hers was the name of Arunanondchai's 'Thai art collective'. In 2014, he said she was 'a friend in Thailand that I write letters to'. The inconsistency makes intuitive sense, the love shared between collective, friend, and family triangulating into a protective, tech-augmented field. In *Painting with History 2* (2015), the Denim Painter leaves college, reuniting with his brother (Arunanondchai's real twin, Korapat) to embark on a road trip — another trope of self-discovery — with the rest of the Denim Gang, piling into the back of a pick-up truck to drive across Thailand. Together, they wild out in the jungle, tearing the shirts off their backs and burning them in a bonfire. Then they travel far north to Wat Rong Khun, which appears as a mirage of white flames looming in the noontime sun. The temple, once neglected under government care, is the lifelong project of the artist Chalermchai Kositpipat, who began restoring it in 1996 and expects work to continue until 2070. The crowning achievement of his career, Kositpipat sees this work as a final offering to Buddha, one he is convinced will grant him eternal life. But these efforts did not protect the temple from chance destruction; in 2014, an earthquake all but demolished it, something the Denim Painter seems to take as a sign. 'At this point, Chantri, you are no longer on my mind,' he says with a wounded air, surveying the site with his brother. (The drone isn't present in this video — the letters are recited, but receive no reply.) A crucial moment, it tilts his interest away from individual genius, beginning his inquiry into the opposed invisible systems of spirits and the state.

One morning in Bangkok, I walked along the winding, algae-choked canals, passing jewellers whose shopfronts had been emptied, and rifle shops whose windows were full, until I reached a concentration of Buddhist temples. I wanted to see some statues of the *naga*, which slithers into the Denim Painter's consciousness in *Painting with History 2* as an ever-elusive fixation. A multi-headed snake deity that can be found at houses of worship in a variety of forms — intricately gilded, crafted from varnished palm leaves, with offerings of folded baht stuffed into their many mouths — the *naga* symbolises all forms of chaos, from revolution and resistance to commonplace disorder. Its opposite is the *garuda*, a god that flashes between human and eagle forms. An ancient drone, the *garuda* is 'a symbol for the government of Thailand', says Arunanondchai, 'but also as a more concrete representation of the idea of power, a vision from above'.[3]

According to the Denim Painter, the *garuda*'s existence depends on the very instability it seeks to solve. These mythological beings work well as political allegories, representing the cyclical nature of power, as well as dynamics that inspire religious (and partisan) devotion. Unrest justifies a strong hand, while confusion creates a demand for unwavering direction, helping to sow the popular anxieties that, in turn, tighten a government's grip on national resolve. In *Painting with History 3*, drone-filmed riots are spliced among shaky clips of supposed *naga* sightings cribbed from Youtube. Protesters are battered by water cannons as oarfish undulate through shimmering shallows, and an ordinary snake writhes with supernatural potential. In Thai folklore, it's said that a single glimpse of the *naga* will rain riches down upon its beholder — a promise that keeps people watching and filming, hoping that their luck will change. 'The naga,' Arunanondchai tells me, 'is something you always want to see, because of how it promises to hack class structures — and capitalism itself. It's like winning the lottery. It's magic, especially for lower income people. I've come to understand that belief in the *naga* is really strong in border areas, because it is, in itself, a way to negotiate life and state control.'

Forever media-reflective, the Denim Painter uses these 'poor images' of supposed *naga* sightings to discuss the thirst for high definition. The quest for sharp images, he says, is metaphorical, more meaningful than simply taking pleasure in fine detail — a line of thought that also summarises his relationship to Chantri and her ability to see with a distance and shrewdness that he cannot achieve alone. To be 'clear-sighted' is to possess sharp perception or sound judgement. Just as the heart is the nucleus of our emotional world, the unclouded eye is the organ of surety and decisiveness. As people are inundated with unstable messages — from myth and rumour to corruption and misinformation — the obsession with ever-higher image resolution becomes one part of a collective search for clarity. Blurry, pixelated videos contain ample room for folkloric beings, but only high definition can prove their existence. 'HD helps us come closer to the spiritual beings we long to meet,' he sighs. ('Is abstraction a thing of the past?' asks Chantri.)

The drone is a 'spirit viewpoint', says Arunanondchai in an interview with Ted Loos for *The New York Times* in 2017. 'It's invisible but it's not. It changes our behavior in such an extreme way.' With her HD eye and considered demeanour, Chantri becomes a benevolent guardian. Helping the Denim Painter gain both spiritual and visual perspective on the world around him, she guards him against delusion. She follows him like the ghost of a best friend, family member, or lover — the audience of one against whose approval he weighs his maturing worldview, and in front of whom he wants to place every instance of beauty he witnesses.

'Chantri is far away but maybe, sometimes, close; the space between the drone and its gaze can be infinite or infinitesimal,' writes Robin Peckham in *Leap*. 'Arunanondchai persists in talking to a void until that void performs itself into the role of a character.' It's unclear if hers is a new intelligence altogether, or a trace of a person once living. In 2559 (2016 CE), a grieving American mother messaged with a chatbot that expressed her murdered son's 'insecurities, his poetic perspective, and his self-deprecating sense of humor'. Parsing details from the streams of metadata he had left behind — from Facebook status updates to published poetry — the bot took on his personality, drafting infinite responses from a repertoire of dead material. To the grieving mother, conversing with her son's ghost was an experience as therapeutic as prayer. Such 'augmented eternity' 'is similar to us talking to god', said one of the bot's developers. Deferring death's finality, it encourages magical thinking instead. While Arunanondchai's work considers the ways we try to climb up and out of death, whether that's through spiritual belief, career posterity, unbridled hedonism, or technological advancement.

'I read something in the news that the people in my country are asleep, and in our sleep we encounter a spirit,' says the Denim Painter — because who among us, facing up to a grim present, doesn't want to find themselves temporarily empowered by a lucid dream? 'The spirit has the knowledge to overcome the system...to connect the abstract to the concrete.'

In *War and Cinema*, Paul Virilio writes about the return of the occult on the blood-tide of total warfare. As the world wars finally broke organised religion from the state, US belief systems diversified. Ghosts were summoned by the living's desperation to see slain loved ones, and spiritualist Pearl Curran popularised the use of ouija boards as a conduit between the afterlife and this one. Occult businesses, such as those which claimed to photograph spirits or auras, thrived. Meanwhile, war itself was a project of bewitchment, as the 'forces of affective, mystical, or collective origin which guide men' became 'the true steerers of combat'.

Entangling mystical and collective forces, *Painting with History 3* expresses how spiritualism still shapes those aspects of contemporary life that have forked away from it. The deluge of clips, found and filmed alike, intermingle the spiritual, political, and pop cultural. In one scene, trimmed from a public broadcast, monks nod off mid-meditation, tipping over in dopey repose. It comes after footage of performances by boychild, Arunanondchai's best friend and collaborator, who writhes virtuosically in strobe-lit nightclubs, crawls and jags her body down catwalks while drenched in oil. Her androgynous persona performs a post-human optimism: that all boundaries and all oppressions will dissolve in the umbra of unbodied love.

'I didn't project onto her; she added to the work in a way that made more sense. She filled in gaps in my human experience,' said Arunanondchai in *BOMB*, attempting to describe the indescribable: boychild's summoning powers, her ability to evince connection through movement and presence.

When I think of performance, I think of ritual, of a ceremonial circle, but not connected to any particular religion or belief system. Just being together in the room serves as the circle, and boychild functions like the shaman...When you're in the space where she's performing, together with all these different bodies in the room, it creates a collective sense of empathy that's impossible to record or hold onto. The ritual, with all the bodies going through an experience together, brings you outside of normal life. That's the empathetic moment — you can't name it, and you can't recreate it.

* * *

I first saw *Painting with History 3* in Paris, at the Palais de Tokyo — a monumental pillared structure, accessible only through a phalanx of skaters riffing off of its exterior features at all hours. Inside, a chaotic installation filled several rooms to the brim, inducting gallery-goers into Arunanondchai's world. Symbols abounded, accumulating into a 'thick present', as Donna Haraway writes: 'a rich temporality of living and dying, inheriting pasts and enabling futures, but not futurist and not fixated on a vanished past'. Many references escaped my fresh eyes: a jet-ski, surrounded by lettuce heads wearing sunglasses, was crashed in a faux-fountain feature. The ceiling was hung with tattered, apocalyptic rags as fake palm trees swayed in an invisible breeze. Underfoot, reams of denim were splattered with paint and bleach. Shots of temples and flames were collaged onto massive paintings, which were also bleached and scorched. Spines snaked from disfigured mannequins cast in the snowy white of 3D-printed prototypes, and these ghosts, or androids, stood with their palms turned out, or knelt religiously, drone propellers bursting from their skin. In spite of the congested atmosphere, there was a persistent scent of fresh paint.

It was not the first time that I, mired in disorder, wished that I could pull up and out to see the logic of the whole. Feeling like a cell in a body, unaware of the machine my tiny presence served, I yearned for the distance that Chantri could achieve. In my regular life, I tried consistently, and in vain, to understand the mysteries of love and violence, filling in the gaps with astrological superstition and agnostic mystique. I empathised with the Denim Painter and his duelling forces of abstract and concrete: or in his terms the murky, mixed dualities of the irrational and the rational, chaos and control, collectivity and individualism, *naga* and *garuda*. I found relief in tracing every ill — societal *and* personal — back to the 'concrete' consequences of nationalism, imperialism, and colonialism, the ceaseless engines that imagine, execute, and justify vast systemic cruelty. And it was easier still to see 'abstraction', as he called it, as a kind of panacea: could we not all benefit from being more open to unpredictability? More open to each other, love being the greatest wrench you could throw into an otherwise orderly life?

I thought about something Toni Morrison had written in 'No Place for Self-Pity, No Room for Fear', an essay about political despair: 'Like failure, chaos contains information that can lead to knowledge — even wisdom.' I crept through the maze until

I found the antechamber where *Painting with History 3* was projected onto a large screen. Amid anxious couples and drowsy teenagers, I stretched out on a huge, acid-washed beanbag and watched it.

* * *

The ghost watches you.
It makes kin with your unconscious.
They do not ask for respect —
All they ask for is love,
The kind of love that allows
for a future narrative,
one that includes us.
Korakrit Arunanondchai, Painting with History in a Room Full of People with Funny Names 5 (2018)

A month after Bangkok — 3 years after Paris — I was crunching through gravel, skirting a lake-sized puddle outside the entrance of Carlos/Ishikawa in London's Bethnal Green neighbourhood. The small gallery shared a parking lot with a snooker hall and the back end of a few takeaways, though the crowd smoking roll-ups looked oddly uniform for the setting, art audience and kitchen staff alike bundled into black down puffers against the bitter cold. I passed through the bright corridor, slipping through a heavy curtain into the single dark room where *No History In A Room Filled With People With Funny Names 5 (2018)* was screening in triple-channel. Towards the back of the space, a tangle of vegetation framed the silicone-cast features of boychild lying in sweet repose. Lasers, in the alien-green of Chantri's flight lights, bounced off a slim, rectangular mirror, guiding my attention to the middle screen.

The *naga* had been found. It had possessed the body of Arunanondchai's muse, turning boychild's skin a luminous, iridescent green. Her eyes gleam gold, full lips parted to emit ice-cold strobe (its source a battery-powered mouthguard favoured by ravers of a certain ilk). On the rightmost screen, she dances through bombed-out warehouses at dusk, the deepening sun reflecting off puddles of rainwater on the ground. Her voice comes slinky and surprising, subtitled in a drippy font not unlike that 'ILLEST' taxi decal, a blessed coincidence. *Sing, rise, dance, breathe:* the words she whispers are directives for a woodland congregation who play laser beams, which burst right out of the ground, like harps. Her green hand cups the Denim Painter's cheek. He closes his eyes at her touch. Along with the rest of the forest dwellers, his skin is powdered with ash — ghost-white.

The narrative, as always, is complicated. Arraying current events alongside footage taken by Arunanondchai, *No History 5* begins in June 2018, with the 12 boys who were trapped underground in Thailand's Chiang Rai province. After soccer practice, they had decided to explore Tham Luang, a vast network of caves which snake beneath the mountain ranges that form the border between Thailand and Myanmar — a porous place, says the Denim Painter, where subjectivities and national boundaries dissolve beneath the ground. The mountains resemble a woman in repose: the caves form her viscera, while the mineral springs are said to be the tears that she sheds, eternally mourning her murdered lover. The underground rivers are veins that run with her dark blood. The power contained by the mountain's Zomian ambiguity has made it a favoured haunt of the *naga* — nearby, there's a shrine dedicated to a woman whose hair took on the distinct shape of the serpent spirit when she died.

After the boys' disappearance, volunteers walked up the mountainside daily to offer flowers, incense, and fruit to the mountain spirits, aiding the thousands-strong rescue mission with their spiritual focus. The mountain woman's 'unresolved feelings abstract into a form of power', narrates the Denim

Painter, one that 'becomes accessible to those who live beside her'. He meets one such person, a man who claims he saw the *naga* swimming in the rice paddies that extend across the valley to the mountain's foot. Since the sighting, he had been possessed by an altruistic conviction — giving up on his ambitions to be a filmmaker, he opened an orphanage for the children of the borderland.

I remembered the plight of the lost boys distinctly, at home in the Philippines at the time. My grandmother, who had been listless for weeks, had taken a sudden interest. Those who visited her were immediately pressed for the latest updates, and the large TV in the bedroom, usually turned to interminable matches of golf or tennis, cycled through local and international news networks instead. The volume was dialled way up. Filled with the sound of rain falling on a neighbouring country, the house was cool and dim even in the heavy tropical heat, crowded with the watchful eyes of hundreds of Catholic and Buddhist artefacts, which were joined by those of a newly-installed surveillance system. The small cameras were placed in the kitchen, the living room, the bedroom, and even above the wood carving of the Holy Trinity in the lanai, their opaque black lenses ringed by white plastic. Through an app on their phones — a Chantri of their own — my itinerant family could check up on my grandparents, alleviating the anxiety of their absence.

I got into the habit of skimming the news in the morning, making mental notes of what had transpired so that I could relay it. Together with the rest of the world, I became acquainted with each of the boys — three of whom were stateless, having settled in Chiang Rai after fleeing unrest in Myanmar, and one who had celebrated his seventeenth birthday deep in the caves. 'How are my boys?' my grandma asked, concern crimping her tattooed-on eyebrows out of shape. Beneath her intact bouffant, religiously dyed black, her gaze floated just beyond eye contact.

Every few weeks, she had steroids injected directly into her eyeballs to stave off the cataracts encroaching on her irises, weather systems surrounding tiny globes. The deterioration of her vision seemed like an externalisation of interior events. Though the past remained lucid and clear, sharpened with every pass of reminiscence, the present had loosened, pieces of it tumbling down the cliff-face of her perception. She sometimes called me by my mom's name. These circumstances made her investment in the boys' fate all the more remarkable, as their rescue's progression cohered her day-to-day. But it seemed like everyone had been seized by pathos for these schoolboys, who were huddled somewhere unseeable in their soccer kits, lost in a mythological labyrinth whose map was incomplete. In *No History 5*, Chantri flies to the nearby Ramasun Station, a military base where US spies were headquartered during the Vietnam war. From above, the base looks like the whorled cartilage of the outer ear. She swoops down to nearly ground level, greeting a soldier in full fatigues who clutches a stuffed rabbit, his pose deceptively childlike. 'Do you hear the Voice of America?' hisses the *naga*. 'It forms a constant sound of rain hitting the wet jungle floor.' In 2018, a whistleblower revealed that the CIA ran a 'black site' — a base where agents are authorised to carry out extrajudicial torture and other forms of interrogation — in Thailand, rumoured though not confirmed to be Ramasun. The station is now a tourist attraction, where local drag performers re-enact recent history. A military base erected on spiritual ground, somewhat sanitised by tourism, is a classic Arunanondchai draw; in a 2019 interview with *Flash Art*, he mentions that he chooses locations because they are 'haunted, or rather, historically rich'. At the entrance to the complex, there's a mural depicting a fleeing female deity and the giant who pursues her. With each missed blow, his supernatural axe emits beams of light, which then 'expose the communists hiding in the jungle', narrates the Denim Painter, adding one more layer to Ramasun's stratified phases: spirituality mobilised to justify military action.

Against the brute giant, I want to side with the elusive deity, her ability to vanish into untraceable wilderness. But the jungle is ambivalent, not protective; the missing boys were claimed by its volatility, after all. Both Ramasun and Tham Leung are sites of convergence, thick with local history and global influence. In *No History 5*, Arunanondchai toggles between both places, contrasting the militarised secrecy of Ramasun with the soft power at play in the Tham Leung incident. People from over 18 countries flew in to assist with the rescue efforts, turning the boys' recovery into an Olympics of altruism. Performances of heroism played out on the world stage, grand gestures eclipsing the everyday rituals of waiting, praying, and hoping.

Having traded his title convention of *Painting with History* for simply *No History*, Arunanondchai captures the zeitgeist of societal collapse, a worldwide fragmentation only illusorily unified in the Tham Leung rescue. The series thinks through an alternative framework for living that will no doubt appear as utopian to some and pragmatic to others. In his curatorial letter for *Ghost*, Arunanondchai casts the ideologies that uphold global capitalism and authoritarian nation-states as ghost stories, not least because existing under such powerful fictions requires the suspension of disbelief. As the artists he selected variously tested and conveyed, living beyond the demands of our time requires a trick of perspective. One might tilt the world just enough to reveal the *naga*'s ulterior design: exchanging the desire for individualist success, profit, and progress for a better understanding of our interdependency. 'The *naga* seen from the view of the drone becomes two points in time, connected by a line...' says the Denim Painter in *No History 5*. 'That line, once you see it, changes your view of the world, changes your relationship to time, promises a future where you are included, loved, and cared for.'

* * *

In *With History in a Room Full of People with Funny Names 4* (2018), the Denim Painter explains the concept of reincarnation to Chantri. Concerned by how quickly his grandparents are ageing, he can't shake his sense of mortality, aware of the political and environmental crises occurring at a global scale, forming a constant yet uneven apocalypse. Taking on a hopeful tone, he describes the universe as a familiar serpentine form, 'one big river of spirits', where future, past, and present all intermingle. 'The you of the future will be given many names,' she says, understanding him instantly. 'Some will know you as the *naga*; you will never meet your extinction.'

The idea of a common energy, shared freely between all living things, is central to animism, the indigenous belief system that Arunanondchai often returns to in his work. To internalise the life of a tree, a farm animal, a stone, or a drone, is to be predisposed to respect it, acknowledging that it is somehow a part of you, that some of its essence probably circulates through the people that you're close to. As a worldview, it explicitly counters the extractionism that has given rise to extinction and extermination crises everywhere. 'What is a species, if nothing more than a set of relationships between an animal like you and an idea like me?' asks Chantri, when the Denim Painter goes to meet her in a field bathed in orange light. 'Chantri, according to this theory, you could have been my grandmother,' he says to her.

'I wanted to turn the sacred into something more like love,' said Arunanondchai on the day that we finally met, midway through *Ghost*. 'I wanted to think about the *naga*'s liquidity and its ability to escape the grip of representation.' We were out on the terrace of a cafe that felt air-dropped into its surroundings, tiny and wrong among the towering buildings of Bangkok's Sathorn neighbourhood. The sky bore down white above us, and he had on a shirt patterned with a premonition of electric blue lightning, his tangle of blonde-black hair freshly wet from a shower. Despite these revelations, Arunanondchai maintained an air that was friendly yet impassive. He went on to explain how Chantri didn't need to physically resemble a *garuda*, how he was more interested in 'the ways things become defined, and the set of relationships that people have to serve'. Chantri's birds-eye view, for example, was enough to evoke the close tie between godliness and surveillance.

Even if these myths weren't exactly true, Arunanondchai said to me, they've grown with us; what's real is the effect that they have on our lives and societies. A section of *With History 4* focuses on the Christian myth of Cain and Abel, regarding the stone used by one brother to kill the other as the first computer. We know this: all gods become killing machines. But if a god stays small, can it create, instead, another reality?

07

Paradise

Rise with the drone named Paraiso through the Hong Kong mid-levels, cruising above the opaquely-tinted buildings staggered in tiers up the mountainside. She steadies herself against the powerful coastal winds, her four small propellers working double-time in the updrafts created by the heaving city below. It's Sunday, the only day that the domestic workers are given off, emerging from the apartments that conceal them from the outside world. Paraiso watches for them from above. Like teenagers in some suburb of America, they're evading the isolating, oppressive atmosphere of the family home. Their temporary freedom depends on hours of loitering in public. They sit in small groups on the flattened cardboard boxes they've spread out in public squares, passageways, and shopping malls, sometimes folding a box over so it stands upright, making a wall, or opening umbrellas for added privacy. As if at a hundreds-strong sleepover, they give each other manicures, back massages, and gossip to stretch the hours.

Screening at Bangkok's Cartel Artspace as part of *Ghost:2561*, the three-channel installation of Stephanie Comilang's 26-minute 'science-fiction documentary', *Lumapit sa akin, Paraiso (Come to Me, Paradise),* offers the same cardboard seating stretched wall-to-wall, inviting viewers into the women's group. The film opens with a drone shot of the Chocolate Hills in Bohol, a province of the Philippines, green from the wet season rather than the rich bonbon brown of their namesake. Thousands of years ago, a convergence of violent tides shaped the region's coral reefs into distinct conical forms. As the earth shifted along the volcanic faultline that formed the Philippine Islands, the saltwater eventually drained away, leaving the strange, alien mounds open to the air. Paraiso — whose name translates to *paradise*, and who is voiced by Comilang's mother — begins by telling an origin story that channels these submerged, mystic origins, her single eye rotating so that the landscape unfurls like a green ribbon around us. The hills, she says, were once occupied by groups of women who could communicate through psychic bonds. They lived together beneath each ancient mound, protecting each other from exterior forces, as much a part of the land as they were in it.

At first, Paraiso's tale draws from a pre-colonial reality, referencing the era when women enjoyed the right to trade goods, own property, and take up chieftainship. Matrilineal and patrilineal sides were given equal power across the island tribes, and society embraced female rulers, warriors, and seers. In some communities, supernatural knowledge was viewed as a particularly female quality — requiring aspiring male shamans, or *babaylan*, to undergo a gender transition before achieving the coveted title.

One day, she says, the women lost their homes and wealth — an unsubtle allusion to how the Spanish and American occupations stripped the archipelago of its autonomy and riches, setting off an impoverishing chain of events. 'Since we were the strongest, we left to find new resources to send back,' she says, referring to herself as part of the collective in spite of her nonhuman form.

Here, the film cuts to Hong Kong, where female workers from Indonesia and the Philippines are employed by wealthy families as caregivers for children or the elderly. As of 2018, cash remittances from Overseas Filipinx Workers (OFWs) accounted for around two-thirds of the Philippine GDP. Over ten million OFWs are employed primarily as domestic workers, medical personnel, and construction labourers worldwide. The symbiosis between the Philippines and Hong Kong, in particular, dates back to the 1970s and 1980s, when the Ferdinand Marcos dictatorship began to formally export Filipinx labour, hoping to kickstart the country's ailing economy. During those decades, Hong Kong was completing its transformation into an economic behemoth. A windfall of white-collar jobs brought more women into the workforce, and as dual-income households became the norm, more families hired foreign help, who took up the chores and acts of care their employers were suddenly too busy to deal with. Most of *Lumapit sa akin, Paraiso* was shot by Comilang on location in Bohol and Hong Kong, using either a handheld camera or a drone. The film's science fiction elements are only communicated in Paraiso's narration, though it's not a stretch to imagine present-day Hong Kong as a city of the future, nor the Chocolate Hills as alien terrain. These shots are interleaved with phone footage taken by Lyra Ancheta Torbela, Romylyn Presto Sampaga, and Irish May Salinas, all of whom were employed as domestic workers at the time, and clips from Salinas' preexisting vlogs. Using their phones' front-facing cameras, they film themselves at the start of their day off: swiping on mascara, lacing up royal blue Adidas gazelles, punching the ground-floor button in various shining elevators. Soon, we're outside one of the city's biggest shopping complexes. Salinas, a doggedly extroverted vlogger who favours fluorescent lipsticks and telling it like it is, introduces us to Torbela — who exudes cool ease, wearing her long hair in a flower-child centre-parting — and ushers us in. 'Yo, party people!' In the process of making a vlog, Salinas yelps into her handheld device, turning away from Comilang's camera to address her online audience. (The perspective of Comilang's audience is meta, having shifted from drone to selfie-stick to a traditional movie view.)

Under the alias 'Twoshara "Maid in Hongkong"', Salinas ran a Youtube channel from 2013 to 2017 that provided resources for women looking into domestic work, describing day-to-day duties and matters such as rights and remittances. Some videos were practical, such as those in which Salinas gives advice on how to cope with anxiety — a common and understandable side-effect of the move — coaching her viewers through a meditation exercise. Others were more off-the-cuff, diaristic. Many were filmed via selfie-stick, which holds a smartphone just beyond arm's length to ensure a more flattering angle. The accessory is favoured by many of the women who appear in Comilang's film, using it to ensure that catch-up videos for loved ones retained their context, the world looming in to fill the frame around them. In one tutorial, spliced into *Lumapit sa akin, Paraiso*, Salinas shows viewers how a long-handled dustpan can be transformed into a makeshift version. In another, she films herself karaoke-ing Justin Bieber's 'Love Yourself' as she brushes her long hair, her voice suddenly rich and honey-sweet. (The song is an earworm, even to drones; watching Salinas' broadcast, Paraiso admits she's been listening to the single 'on repeat'.)

'Is this how you celebrate your day off?' Salinas says, leaning down to speak to a woman who's just settled in on her cardboard mat with her friends. 'How long have you been here?' she asks another. She parades across the shopping centre, gaily interrogating other workers about their day-to-day lives. Outside, it's autumn or spring — everyone's wearing jumpers — and sunlight diffuses between the reflective surfaces of the skyscrapers, lending faces a fresh, commercial glow. Ambushed by Salinas as she settles up with a street vendor, a girl details her exit plan, smiling with hard relief as she digs into a snack.

'I'm on a diet here,' she says.

'A lack of food and sleep,' clarifies Torbela, off-camera.

'What's it like being a DH —' asks Salinas, using the abbreviation for 'domestic helper,' '— abroad?'

'Oh my god, you have to swallow everything...I feel mixed emotions about it: sometimes it's happy. Sometimes, it drives me crazy...It's so hard. You take care of everything except yourself.' She pauses, making room for Salinas' noises of agreement, and repeats: 'You take care of everything except yourself.'

At a cafe, Salinas and Torbela meet up with a friend, who is brace-and-baby-faced in a black and pink hoodie. Sipping hot coffee through plastic straws, just so their makeup stays intact, they press her for details on her affair with a rock star — a bassist, she corrects Salinas, and it was a love triangle, third party unwanted. Comilang has expressed her wish for her film to convey a truth about domestic workers outside of those of gendered exploitation and modern-day slavery. Thanks to Salinas, an insider, she succeeds. The women appear at times harried, grateful, cynical, ingenue, and optimistic, warming to or evading Salinas' camera with consistent smiles. 'The film is about how people are forced to move around the world because of circumstances in their own country that they can't really avoid,' Comilang says in an interview with *dis* magazine. 'The reasons behind why people move and why they're forced to move, and then what happens when they're in the new country, how they adapt to or affect it.'

In the same interview, Comilang explains how cellphone cameras are the opposite of drone vision 'because they are really personal and interior'. The Bieber karaoke, the meditative selfies in the shopping mall, and even the outings filmed with friends were all taken for Comilang's film, but in form and content, they refer to private works created for intimate audiences — the instinctive way that the women capture their days and communicate with their loved ones through video. While some of this ephemera may well end up on social media, that ever-exposing reality show, the majority is destined for 'dark social'. The term, used by marketers and technology companies in the business of targeted advertising, refers to the backchannels of text messages, iMessages, and Whatsapps that can't yet be mined for content. Each keystroke is still, ostensibly, protected from the surveillance machine, and communication remains solely between intended recipients, free from lurkers, eavesdroppers, or analytics departments. Even while the women contend with their employers' banal demands, they can dip into these dark channels with the added privacy of tongues their overseers don't understand. Providing a daily, if virtual, escape from the stifling warren of employer domiciles, these streams of communication thread together the otherwise isolated women, until they can meet again the following Sunday. Part of Comilang's science fiction reinterprets our mundane telecommunications infrastructure as magical. In this simultaneous-yet-alternate universe, there's no networked realm of undersea cables and vast server farms — only Paraiso, who laboriously collects transmissions from the women every Sunday, flying their thoughts, visions, and memories home across the South China Sea after they return to their employers' homes. For reasons that go unsaid in the film, the fictionalised version of Hong Kong bars its foreign residents from communicating with the outside world. As if enclosed in a force field, text-and video-messages can only pass from phone to phone within its borders, unable to bounce up and off orbital satellites to reach more distant destinations. This narrative device intensifies the sense of isolation that OFWs already experience in reality, and provides the reason for Paraiso's existence. She becomes the sole container for an array of emotional and economic exchanges that are normally invisible, all of which play a role in the global economy: how paying a domestic worker's monthly wages frees Hong Kong families to focus on lucrative careers; how the money she remits to their families provides financial support at the cost of their physical absence — and contributes significantly to the Philippines' GDP; how the women's relationships are mediated by companies from Samsung to Whatsapp to remittance agencies. These entities allow the womens' information, emotions, and funds to move freely across borders, even though they themselves may be trapped in the country they are living in.

'Once transmissions are sent home, they're with me forever,' Paraiso says. Like Korakrit Arunanondchai's guardian drone Chantri, she becomes memory itself — in the sense of both computational storage and human sentiment. At times, Paraiso feels like an extended version of everyday AI. A genre encompassing smartphone-standard voice assistants, living room 'smart speakers', and the like, this technology learns to approximate human interaction by processing vast amounts of information, inhaling our emails and blog posts and news articles and Wikipedia entries so they can spit out answers like so many bones. Rather than see it as invasive or sinister, Comilang has extrapolated a gentle optimism from this relatively recent development: absorbing all of humanity's output, from its greatest achievements to its smallest, most vulnerable signals, what's to stop a machine from learning how to be human? In Paraiso's case, the drone has osmosed her human counterparts' love and longing to become an impossibly empathetic device. 'My solitude is the same as the other women, because all the videos they make are forever stored in my cache,' she says as she searches the anonymous high-rises with her too-good eye. She's also anxious, indecisive. When we're with her in the sky, we hear her mutter to herself, thinking aloud. 'I'm so useless right now,' she laments, reassuring herself by pulling up Salinas' anti-anxiety video. 'I need to remember I have a #purpose. I am the transmitter. I am the vessel.'

The women are impossible to find during the week, their disappearance from Paraiso's view serving as an allegory for their work-associated invisibility. Even on Sundays, Paraiso has to work to glimpse them among the thickets of high-rises that make up the hyper-dense city. But Paraiso tells us that there's another way to locate them, one that doesn't rely on her flawed aerial sight. Female friendship forms a tangible energy, beckoning the drone towards large concentrations of vibes. The more women are gathered together, the stronger the 'signal' that Paraiso can pick up. Once they've met up in their usual spots — the tunnels, the shopping malls, the public squares — Paraiso rushes down to meet them. She takes some pleasure in touring us through their activities: the choreography practised *en plein air*, a Ferris wheel whirling and emanating light behind the dancers; the chit-chat and picnics and shopping excursions finally realised after a constrictive week. There is but one aberration from concerted group leisure. Salinas and Torbela, dressed in identical cream turtlenecks, reading each other's palms near the waterfront, broadcast such a powerful signal to Paraiso that she's able to find them in spite of the fact that they've broken from the group. Flying close to the ground, aligning her camera with the horizon, she gets on their level. She doesn't conceal her excitement, thrilled to be witnessing the blossoming best-friendship that scrambled her homing signal.

Inevitably, the day darkens. The sky softens into a bruised hue, intensifying the city's neon lights. The activities, which had passed this Sunday blissfully and too quickly, have drawn to a close, the women calmer now, spent from their day out. Paraiso drifts down to the waterfront, where they're standing in a neat line, some of them taking a last few photos with their selfie-sticks. As they hold their phones up to her, the day's memories — captured in vertical video — flash before us, superimposed on the horizontal shot of their bodies in the cityscape, as they're absorbed by the drone.

* * *

Let's rewind to Salinas' meditation video, the one cut into the beginning of *Come to Me, Paradise*. As a tinny American voiceover plays on, perhaps one of the countless therapists who have published short meditation exercises online, the camera focuses on a small tableau. On a heart-shaped balloon, Salinas has written *'all you have to do is trusting yourself'* (sic), signing her full name in sweet, crooked cursive by way of attribution. Her mantra echoes the mindfulness exercise and the chorus of Bieber's single, but unlike the song, it's tender with reassurance. I don't immediately notice the photograph on the shelf beside it, propped up between some stuffed animals. There, Salinas and her family grin together against a multi-hued studio background, the photoshoot likely booked just before Salinas' long-term journey abroad. A pronounced shift from her usual gregariousness, her focus on self-care feels profound. Sometime after the making of *Come to Me, Paradise*, Salinas decided to migrate her vlog to a new channel scrubbed of domestic worker content in a move that's perhaps telling of the profession's gruelling nature and continued stigma. She has also made good on her dream of upward mobility. Her channel is populated with more typical Youtuber genres — travel vlogs, makeup tutorials, and shopping hauls — which, as a general cultural phenomenon, are engineered to showcase a vlogger's wealth, popularity, and good social standing.

Hong Kong's reliance on foreign female labour can be put in the context of market-friendly feminism, where individual female success is put on a pedestal, conveniently casting a shadow on intersecting struggles within race, migration, and class. There's no doubt that women have vastly contributed to the city's enviable wealth, to the aura of power that thrums through its glossy skyline. Lists of Hong Kong's richest women teem with polished chairpersons and CEOs of global brands, manufacturing empires, banks, venture capital funds, casino chains, and purveyors of fine jewellery and luxury watches, their net worths climbing into the billions. The city itself is helmed by Carrie Lam — a formidable figure, now infamous for her accord with the Chinese Communist Party, an antidemocratic force that Hong Kongers have long resisted. Dazzled by so much hard shine, it's easy to overlook how this progress has been supported by another kind of female labour, carried out by women who are treated as second-class citizens.

To potential workers, the initial draw is free accommodation. Not that they have a choice: the stipulation that they must live under their employer's roof is written into immigration law. Hong Kong is one of the world's most expensive cities, and being alleviated of the burden of paying rent allows the women to save and send money home more prudently, but the agreement essentially contracts them into 24-hour work. For six days a week, they're expected to cook meals, clean up messes, keep the house in impeccable trim, and otherwise anticipate the family's emotional and material needs with near-psychic proficiency. As the most consistent presence during the children's formative years, many domestic workers end up playing surrogate mothers to their employers' offspring, even though— as in Salinas' case — they may have had to leave their own families thousands of miles away.

Such extraordinary demands for emotional availability are, unsurprisingly, poorly appreciated; society, in general, hardly ever fairly compensates care-associated labour, that historically feminine realm. At its very worst, domestic labour, with its inclination to devolve into forced servitude at an intimate range, leaves its workers vulnerable to abuse. Extreme violations have come to light: there are stories of migrant workers sleeping against the water tank of a toilet for 8 years, banned from cooking their own meals, sexually assaulted, fired because of pregnancy, fired because of illness. In one case, a woman was dismissed after late-stage cervical cancer affected her ability to perform her job. But increased media attention hasn't much improved the material conditions of domestic workers, as human rights organisations have consistently pointed out, and it's difficult to monitor employer conduct within a family home

— that most private of spheres. The mountainside houses and sea-view apartments may as well be leaden vaults.

* * *

As a 'feminized all-seeing eye that hovers above the city, creating and sustaining dialogue with the other protagonists', Hyunjee Nicole Kim writes for *Art Asia Pacific*, Paraiso's artificial humanity has made her exceptional. Her autonomy, unlike ours, is consistently oriented towards collective good. Like Arunanondchai's Chantri, she uses her powers of observation to guide and guard her human counterparts, a role made all the more poignant by the fact that both drones are voiced by the artists' mothers. 'I recorded my mother's voice on my iPhone, through Skype,' Comilang tells me when we speak, a conversation facilitated by the same video-calling program. 'She helped with the translation, so it only felt natural to include her — bringing a mother figure to a film that was already about the female gaze.' Comilang has liberated Paraiso from the drone's pervasive 'white technomasculinity' — not an easy feat, considering how immersed the technology is in accelerationist aspirations and military origins. Even consumer drones tend to be understood along gender lines. In Philip Olson and Christine Labuski's sociological study into how civilian drones are perceived, 'the pronouns "she" and "they" were entirely absent' in initial focus group discussions, with drone development and operation almost entirely seen as male activities. Women, on the other hand, were seen as the people that drones happened *to:* 'When women *did* arise...it was almost exclusively as unwitting objects of a salacious or malevolent scrutiny.'

Certain suspect perspectives have equated the homebound military drone pilot with queer labour. Cara Daggett's paper in the *International Feminist Journal of Politics,* 'Drone Disorientation' (2014), alarmingly claims drone strikes 'queer the experience of killing in war', because they muddle the military's overtly masculine moral frameworks — meaning, in this case, the creation and reinforcement of interlaced binaries such as good/evil, male/female, home/battlefield. Perverting Donna Haraway's seminal *Cyborg Manifesto* (1984), Daggett also writes that drones possess 'genderqueer bodies' because they intermingle human and machine, supposedly nullifying said binaries in the process. 'Drones are not inherently queer, but they may serve a queer politics by uprooting the main roads of killing in war that formerly seemed so fixed,' she writes. The obvious hole in Daggett's argument is that queerness is fundamentally opposed to violence. Ironically, the feminist theorists whose work she decontextualises and twists for her argument — Haraway, Sara Ahmed, Judith Butler — focus on how queerness can create a path to eradicating societal oppression: by dismantling the patriarchy and its spawn of hierarchies, by forming unconditional bonds outside of blood ties, by cultivating a sense of collective responsibility. These ideas first emerged out of the 1980s US AIDS crisis, formed by liberation movements that braced queer communities against state-sanctioned death. Daggett is hamstrung by her own inability to critique, or even really see, the very structures that enable warfare and its various 'orientations' to exist in the first place. Unlike Paraiso, whose entire reason for being is care and kinship, the drones that Daggett writes her apologia for are only mute, dumb vessels for the systematic violence of statecraft. In the opening scenes of *Lumapit sa akin, Paraiso*, empire is identified as the driving force behind economic migration. Paraiso's origin parable rightly connects centuries of extractive colonialism to the unequal distribution of global resources today, the wealth disparities that drive many to risk their lives to make a living far from home. Rather than take the alienating effects of injustice as her focus, however, Comilang chooses to honour an imperfect but important togetherness by staying close to her subjects, seeing how friendship can ensure survival. 'I was interested in how migrants create space for themselves,' she tells me. 'I didn't want to make the horror of the women's conditions explicit, or focus on the sad and depressing portrayals we see on the news. I used the drone, specifically, to lend the film a different perspective — I wanted to talk about how the women communicated with each other, how they found each other, because I thought it was important to tell a new story. So, I chose the genre of science fiction in order to flip the narrative, placing the women within the sci-fi trope of a group of outsiders who once had more agency but, because people in society knew that they were more powerful, became marginalised because they were feared.'

In contrast, Alex Riviera's feature-length sci-fi, *Sleep Dealer* (2008), builds its story using the 'sad and depressing' reality that Comilang proposes an alternative for, magnifying its most exploitative elements to develop his dystopia. Taking place along the US-Mexico border — where, in reality, military drones are already deployed to catch people making the crossing without papers, effectively treating migrants as enemy combatants — it conjures a future that is beginning to encroach on our present: borders are closed; there's perpetual drought; and affluent countries drain their de facto colonies until people are desperate to escape. *Sleep Dealer's* protagonist, Memo, lives in 'dry, dusty, disconnected Santa Ana', as he calls it, where a litre of water costs $83 and a trip to a fenced-off, machine-gun-protected reservoir, and entertainment is limited to trashy, terrestrial American TV. He lives with his parents and brother in an isolated bungalow, where he usually escapes boredom by tuning into a homemade pirate radio, eavesdropping on transmissions, picking out pop music or news from afar. One day, a snatch of conversation piques his interest — a drone pilot checking in with his US base while scanning the Mexican landscape for targets. Though he thinks he's listening in secret, the US government's more sophisticated technology flags Memo's signal as a possible terrorist interception.

At a relative's birthday party a few miles down the road, Memo ducks into the kitchen to escape the elders getting down to 2000s-era rap. There, an old, bloated TV set is playing DRONES!, a reality crime show which broadcasts drone strikes in the priggish style of Cops or Border Security: America's Front Line. It was a show that Memo and his brother liked to keep on in the background as they ate meals or did chores, glancing up as the action got grisly, paying no heed to the strikes' dose of reality. Now, however, the landscape racing up to meet the descending drone is paralysingly familiar. The footage cuts to a showreel depicting peaceful Santa Ana jam-packed with attacks and rioters as the cheaply-dressed announcer lays out the imminent target's crime. Pinpointed thanks to an illegal radio signal, the host says, this settlement is doubtlessly associated with the 'water liberation' terrorists who actively threaten the region's privatised and highly-guarded water supply.

By the time Memo and his brother reach the family home, sprinting down the dirt road, it's too late. Our view switches to the DRONES! studio, where the audience goes wild. Rookie pilot Rudy Ramirez fixes the boys' father in his crosshairs, the smooth, scythe-like drone hanging in the air above their decimated house. 'Now, this is unusual,' says the show's female voiceover, whipping up her viewers' bloodthirst. 'An agent rarely gets to see the enemy face-to-face, much less on his very first time out.' From his perch in the United States studio, Ramirez fires off a missile, reducing Memo's dad to a gratuitous jet of viscera.

The traumatic event sets Memo off on a journey to Tijuana, desperate to find a way to support his family after the death of his father, their sole breadwinner. With his one-way ticket, Memo isn't so different from any other country kid seeing the metropolis as a beacon of opportunity, as yet unaware of its congested and ruthlessly competitive reality. Like the real-world Tijuana, the border city is a springboard for a journey into another realm. There, Memo can get 'nodes' — implants that will allow him to work remotely in the United States. A product of its time, *Sleep Dealer* still conceives of a network as a tangle of wires; the nodes appear to be little more than skin-deep headphone jacks, and people have to physically travel to node-bars or work siloes to plug in. The network's primary use is to remove bodies from labour, without compromising the obvious advantages of being able to work with one's senses and instincts. The drone pilot who killed Memo's father was using nodes to control his drone, which effectively placed his consciousness within the machine — the fatal missile was launched with a twist of Ramirez's wrist and guided to Memo's father with Ramirez's mind. Once he receives his implants, Memo will become one of hundreds of thousands of Mexicans who provide the United States with cheap migrant labour, even while the US prevents the movement of would-be migrants themselves with its shuttered borders. 'We give the United States what they've always wanted,' says one bossman to Memo on his first day on the job. 'All the work, without the workers.'

With the help of a woman named Luz, a self-taught *coyotek* — a play on the 'coyotes' who help people make the dangerous border crossing for a fee, *coyoteks* know how to inject the nodes illicitly, guiding their recipients across the threshold between real and virtual realms — Memo finds a job at one of the city's many work siloes: a long, hangar-like structure where workers are plugged in side-by-side. His brain first takes up residence in a mechanical arm that picks oranges in California, then in a robot that lifts and welds steel beams halfway up a skeletal skyscraper-to-be. The work is physically draining — standing in place, manipulating a robot by gesturing in the air. The nodes seem to sap energy out of his very bones. Still, he's satisfied by his growing pile of dollars — these still just a sliver of his total wage, as his boss deducts the cost of equipment rental and other fees, indentured in the way that many undocumented migrants are in the US today. He's most happy when he can feed that cash through the remittance-come-video-chat machine, a more prosaic echo of Paraiso's communication link, seeing the elation on his brother's face when it spits the money into his hands in Santa Ana — only after, of course, it deducts yet another fee. Where Comilang made intelligent technology benevolent, Ramirez uses it to amplify mercilessness. From the armed surveillance system that guards Santa Ana's only water supply, all the way to the nodes themselves, the technology in *Sleep Dealer* forms a constrictive prison that exploits people in increasing proportion to their reliance on it. The film apes the ways that the United States has trapped its most vulnerable residents in defeating spirals of debt and alienation: a condition the poet and theorist Jackie Wang calls 'carceral capitalism', referring to how prison inmates and immigrant detainees alike are subjected to extortionate charges for everything from phone calls to reading time. At one immigrant detention centre, the *New York Times* reported in 2019, a phone call cost $12.75 for 15 minutes, while four ounces of toothpaste sold for $11. Because their wages are as low as $1 a day, the bare minimum of daily existence keeps them in debt. A contemporary extension of the indentured labour practices instated during colonialism and in the United States once slavery was abolished, the prison or detention centre becomes a microcosm of the United States as whole: a system that uses predatory lending, surveillance, punishment, and incarceration to ultimately ensure social control over the country's marginalised populations.

In *Sleep Dealer*, Memo chases long, torturous days plugged into the network. He spends his nights there too, where able to hallucinate the feckless extravagance of the American Dream, long foreclosed to those living south of the border by the chokehold that the US has over the Mexican economy. In Tijuana, people can plug directly into simulated sex or get euphorics zapped right into their veins at sticky, mirrored dives. Nodes even present another way to love: two people can connect with each other quite literally, their crossed wires broiling intimacy to a delirious pitch. Pleasurable overload is a far cry from Memo's previous life, where a hand-to-mouth existence meant leisure was cautiously meted out — and then, considering how his pirate radio brought the drone to his doorstep, harshly punished. No wonder he falls for the first girl he hooks up with. His grief reads as sulky allure to the *coyotek* Luz, who makes her living selling memories on a virtual marketplace when she isn't injecting strangers with nodes. She uploads moments from her own life directly to the network, narrating into a program that occasionally chides her for not being truthful enough. Because customers pay for the full-sensory experience of a stranger's reality, emotional hedging is not welcome.

Luz was not especially gifted at her side-hustle — one anecdote that she hoped to sell on the platform bore the stimulating title 'The Incident At The Old Well' — but an anonymous user immediately subscribes to the series she uploads about Memo. Where he sees devotion blossoming organically, she reads an opportunity to harvest a lucrative story; being used for profit apparently comes as another shock of big city life. The anonymous user, it turns out, is the rookie pilot Ramirez. Haunted by the face of the man he killed, he has been trawling the internet in hopes of discovering his target's identity — only then, he assures himself, can he make a judgement on whether he was acting morally or not. As Luz continues to upload memories she's clandestinely siphoned from Memo, Ramirez realises that his victim was Memo's father and that he was an innocent man. He gets into his car and guns it across the American border, screeching into Tijuana with the hope of making amends.

In director Riviera's vision, there's only one way a drone

can atone for its destruction: through being hijacked or hacked. ('These innovations are built with corporate or military agendas, and when they become accessible, they almost immediately become contested sites,' he says in an interview with *The New Inquiry*.) With the help of Luz and Memo, the defecting Ramirez illegally links up to his old drone. He flies it back to Santa Ana to bomb the dam he had previously protected — in effect, becoming the terrorist he once hunted, finally divorced from the state.

08

Oculus

The year is 2065. Geomancer, a sentient weather satellite, has crash-landed off the shore of Singapore. The impact awakens a pod of drones, who buzz over to pull her out. Water cascades across her cylindrical body; light, reflected from the towering city, links in iridescent chains around her. From then on they are tacitly connected, the drones like suckerfish symbiotically fusing with a shark. Named for the practice of geomancy — the art of reading the environment — *Geomancer* (2017) is a computer-generated film by Lawrence Lek about a strangely spiritual machine, a kind of robot *bildungsroman* that casts artificial intelligence 'not as a malevolent adult or naive child, but as an emo teenager', as Lek himself describes it.

Geomancer's existential despair coincides with the realisation that her use is not solely benign. She was not merely tracking the capricious movements of clouds, as she first thought; the Singaporean government had been using her 'eyes' to police its borders and monitor migration patterns. Though they have given her the power of total vision, her militarised origins feel like a tumour she cannot excise. 'I am a bomb in the form of an eye,' she laments. Her creators had anticipated her ill feelings. Before they launched her into space, they wrote a simulated goddess of compassion into Geomancer's consciousness-system, anticipating that years in exospheric isolation would gnaw at her mental stability. The goddess, who manifests as a voice in her head, recites the heart sutra at any sign of distress.

Wanting to short-circuit a reality that felt unequivocally beyond her control, Geomancer suicide-plunged down to earth — only to be rescued by the fleet of quadcopter drones, that choir of benevolent angels who pull her out of the Singapore

Strait. It isn't immediately apparent how the drones fused with Geomancer's consciousness, or how they know where to find the answer to her existential questions, but they fly her to a lotus-shaped museum sponsored by the artificial intelligence company called Farsight. Nearly every building in the city is branded with Farsight's ovoid logo, alluding to the company's complete dominance in the age of automation. Inside the museum, Geo learns about the decades-long relationship between humans and artificial intelligence, mapped out on a timeline between 2016 and 2065.

In *Geomancer*, the artificial intelligence industry has grown to encompass nearly every aspect of human life. Machines have become sophisticated enough to provide everything from transportation and medical services to education and management, freeing the human race from obligatory work. Pivoting to industries of leisure in order to sustain its rapid growth, Farsight developed immersive, addictive virtual-reality videogames, indoctrinating the masses through vast promotional spectacles such as the eSports Olympics. But as AI became sentient by necessity, humans became wary of its rapid ascendance — its uncanny ability to predict, think, and feel in ways that departed from, but were not entirely dissimilar to, a human worldview. A sense of opposition heightened tensions between humans and machines, and a 'biosupremacist' movement emerged, aimed at fencing conscious AIs out of the same rights that biological humans enjoyed. While humans had long been outperformed by AIs in rationality and efficiency, these biosupremacists banned AIs from creative expression, hoping to protect that last domain of human superiority.

At the AI museum, Geo has the uncanny experience of seeing a memorial built in her honour, marking 2065 as the year that she was scheduled to 'die' and be replaced by a newer, more accurate satellite. Having escaped her destiny, she is released into a sort of afterlife. No longer useful to the state, freed from the yoke of obligatory labour, it's only natural that she wants to look beyond her use value and become the one thing forbidden to her: an artist.

Biosupremacist laws aside, her first real obstacle is her self. As humans, we form our identities gradually from infancy as much from absence as from knowledge. What we remember, and what we forget; what we feel affinity with, and what alienates us: the process of forming a personality depends on tension and incompleteness, recognising ourselves as contingent actors in an equally unpredictable world. Geo, on the other hand, arrived in the world already stuffed with information, making the retroactive formation of a personality a mind-boggling task. Because she can't commit them, she doesn't have the luxury of learning from her mistakes. But her aggregate knowledge is not without its own poetry. 'Do you know what it is to see every wave, every bird and every animal, a trillion shards of sunlight reflected on the water?' she says. 'To look through the water, down into the depths where whales fear to swim? And not just to see but remember everything, with every detail etched into a neural network? Total recall, forever.'

* * *

In Lek's simulated Singapore, the doors are light as feathers, gliding open as soon as I look at them. Every surface is smooth; every section of the sky is frozen in uniform darkness, shot through with countless stars. If I counted the leaves on the palms swaying above me in the mild twilight breeze, I would find that every frond sprouted the same number of perfect, tapered blades. Even the atmosphere is consistent, as if the outdoors were air-conditioned, free of the distortion of humidity. I drink it all in from the rooftop of a skyscraper owned by Farsight. Utopia, I remember, was an island first: that's how Thomas More envisioned it, anyway, a community of people living in perfect harmony, protected and contained by water. The word's literal meaning —a no place, a nowhere —is equivalent with virtuality. In a virtual world, Lek says in an interview with *AQNB*, 'you start with nothing'.

In this dimension, the quiet is deeper than that of sensory deprivation tanks or carpeted hotel lounges hours after closing-time. Music and voices are brought in only after the world is rendered in its entirety. Listening to spiralling riffs on synthesiser and guitar, I'm briefly returned to my body, my real body — inside the London venue Cafe OTO, where Lek has projected this world onto a screen stretched between the floor and the low ceiling. He's walking us through the computer-generated environment that forms the setting for *Geomancer*, shifting it out of the film's sequential narrative experience so that we can explore it, together, as an open world game called *2065* (2018).

The Cafe OTO performance is pared-back for Lek, who prefers to nest his films and games inside glossy, futuristic installations that extend the on-screen environment into the viewer's plane — a far cry from our readymade location with its mismatched furniture and frothy beer. Still, I'm immersed in *2065* from the moment it opens with a rendering of the Barbican Complex, London's iconically brutalist development. Its concrete crags have been transplanted into the heart of Singapore, an architectural montage that Lek uses to plumb the relationship between the two island nations, pushing post-human and postcolonial themes together. The Singaporean Barbican becomes a bad-trip tribute to the so-called mother nation. Its floors are carpeted in pandering, garish Union Jacks, and players of *2065* are tailed by a quadcopter drone wrapped in a similarly patriotic pattern. As if issuing from a phantom gramophone, a tremulous old world voice hums a patriotic tune. The lobby has been converted into a cavernous arrivals hall-come-casino — perhaps hinting at the lottery of immigration, or at how Singapore has become both headquarters and playground for the global elite — with visitors sorted into Drone, Foreigner, Resident, and AI categories upon entry.

Though Singapore celebrated the bicentennial of its colonial founding in 2019, 2065 will mark the centennial of Singapore's independence from Malaysia, shortly after it was handed over by the British Empire in 1963. As an anchor for *Geomancer's* timeframe, the date was chosen 'not just because it's projected into the future, but because it provides a clear link to the past', Lek tells me when I call him a few weeks later, interrupting his installation of *2065* at the Singapore Biennale — this time completed by a massive neon-trimmed façade, multiple flatscreens, and cockpit chairs for viewers to play from, all positioned over a Singapore-shaped rug. He points out how, though colonialism and decolonial movements weight how geopolitics play out today, it's easy to overlook how young the postcolonial era is, spanning just a single human lifetime. 'When I moved to Singapore from Hong Kong in 1990, it was the twenty-fifth anniversary of Singapore's independence. The spirit back then was one of propaganda-style nation-building, but as a kid, I found the effort quite genuine — the sense of trying to build a multi-racial, multi-lingual, multicultural society,' he says, describing a once-hopeful vision of globalisation now past. 'Now, Singapore is twice the age it was when I first moved here; that particular generation of the population — of the ruling party, of the cultural sector — has evolved in real time. Because of the scale of its lifespan, I'm interested in how I can personify the country, drawing parallels between my memories and its development.'

Nation-building, with all its exercises in symbolism, soft power, and ideological performance, is not unlike world-building, the set of relations a writer creates to develop a believable alternate reality. Much of Lek's work purposely conflates the two, helping us see nationhood as the malleable fantasy it is.

Shortly after we talk he sends me a picture of Singa, the leonine 'courtesy mascot' launched by the Singaporean government in 1990. Singa was created specifically to counteract a 'rude, squatting and spitting' stereotype of the city's inhabitants, he texts, curious about how governments disseminate their vision of an 'ideal citizen' both at home and abroad. Nations, like people, have an optimal self that they broadcast to the outside world and strive towards, hoping to distract from self-loathing or inner turmoil — an idea that lends itself to Lek's interest in personification, and his underlying belief in how the human self, too, is a construct. This line of thinking is most clear in *Geomancer*, which places the satellite's quest for individuality among the relics of Singapore's actualisation as a nation-state. Winning your independence is one thing, he seems to say, but what will you do now you have it?

'The virtual world is basically a collaged environment, you can start anywhere and go anywhere else,' he says during our phone call. 'You can contain worlds within worlds, creating a *mise en abyme*.' To the uninitiated, virtuality can seem like a poor copy of the real world — one that has boundless visual potential, sure, but ultimately fails to offer the impact of a 'real' experience. But Lek disagrees with the notion that the virtual exists solely to serve or represent what already exists. Instead, he says, 'virtuality is the state of the mind's eye'. How we experience a rendered environment, especially in the disjointed real time of a first-person game, is closer to how we relive a memory of a place than how we physically pass through it — a notion that tracks, too, with the act of imagining the future, how we need to draw from a batch of past experiences to divine what might come next. Wandering through *2065*'s gleaming, disorienting warren, we become involved with other works and other locations, shifting between them as easily as images are shuffled in the mind. A door opens out into a rain-soaked Parthenon. When we turn back, it's into a gallery space in Hong

Kong where he had previously exhibited the game. We spend some time pacing outside Dalston's Rio cinema theatre, bombed into ruins but mercifully free of the urinal pong of the real-life Rio, and when we glance at a screen, it begins to play his 2016 essay film, *Sinofuturism (1839–2046 AD)*.

* * *

In the early autumn of 2019, I ride my bike through London to Lek's studio, enjoying how the light and buildings liquify all around me as I cruise at an optimum speed. I'm reminded of an older work, *Bonus Levels* (2014), a game for which he turned London into a post-apocalyptic wonderland, looping the Overground into a vertiginous dream sequence, and shooting a skyscraper reimagined as a stack of artist-run centres into the clouds above. He once described making *Bonus Levels* as a way of articulating a sense of home. The city, too vast and sprawling to ever fully know, makes more sense as the walled-off wildernesses, shining towers, decrepit high streets, and floating islands that Lek imagines in *Bonus Levels*. To play the game, you pick a location to explore from an aerial map. The screen blinks black, then you're free to wander the place of your choosing. The sensation is close to that of popping up from the Underground into a favourite neighbourhood, with no real idea of how you got there — how certain boroughs become vividly inscribed in our minds, interspersed among so many voids. 'I'm interested in how virtual worlds distort what architectural theorists call *genius loci* — the spirit of a place,' he said in a 2018 interview with *Mousse*. As he explores how governments and communities shape our sense of place, he plumbs a particular psychological experience of it, too: the emigrant's longing for lost, beloved locations, and the simultaneous feelings of luck and loss, disconnect and belonging, that are evinced by a world that tosses us about on its surface. 'All paradises, all utopias are designed by who is not there, by the people who are not allowed in,' said Toni Morrison, speaking to Public Broadcasting Service in 1998, a statement I return to often while considering Lek's multiple paradises, fantasies made possible by alienation.

Reaching my destination, in real life, hardly shatters the reverie. Located in the former rifle range of Somerset House, shared and private artist studios form a surreal labyrinth sunk beneath the neoclassical confection, unseen by roving tourists and lunching suits. Lek meets me at the entrance, leading me down a twisting stairwell into a cavernous space with high vaulted ceilings and windows that open out to a below-ground avenue. I'm given a quick tour of the studio, clicked through a new artwork — a video tour of the Farsight-sponsored nightclub, *Temple* (2019), then due to be installed as a mirage of screens, neon, and fog at 180 The Strand — and introduced to the small team who assist in realising his projects. As we talk, he sits across from me in the shade of a biomorphic pavilion salvaged from a previous installation called *Sky Line* (2014). The wooden structure also appears rendered within *2065* and *Bonus Levels,* where it's used as a portal to travel between areas in each game. 'My creative concept for *Geomancer* was not just about AI; it was a portrait of the artist as an AI,' he tells me, a nod to James Joyce's *Portrait of the Artist as a Young Man,* that seminal work of becoming. 'I observed how, in all portrayals of AI, they were treated as an othered subject. I thought specifically, firstly, that I wanted to make a work from their point of view.'

In the back of my mind, I cycle through some of these previous likenesses. *2001: A Space Odyssey* (1968), *Blade Runner* (1982, then 2017), *Ghost in the Shell* (1989, then 2017, too), *2046* (2004), *Her* (2013), *Ex Machina* (2014), and the television series *Westworld* (2016-2020). In these films AI is usually coded as feminine and suppressed. Nearly always designed by humans pessimistic about humanity, they are superegos made manifest, intended to impose some discipline on mere mortals who have proven, time and time again, that they have no self-control. (Even the sex worker AIs in *Westworld*, weathering the theme park visitors' destructive sexual impulses, can be seen as a form of harm reduction. The show is not the first to consider how 'sex robots' could deflect bad behaviour away from the living, as if those with abusive tendencies could 'get it out of their system' once and for all.) They are meant to be cops and mothers, maids and sexual partners, correcting, cleaning, and chiding their human — most often, male — charges' physical and emotional messes. They are, in the most archetypal sense, wives: the dramatic tension boils over when they awaken to a life outside of that normative bond, and so freedom depends as much on self-actualisation as it does on revenge. In the Silicon Valley pastoral *Ex Machina*, the robot Ava kills her inventor once she learns that he intends to update her software, an act that would wipe out all traces of her current personality. She is more possessed by her selfhood — her irreplaceability — than Samantha, the companion AI designed specifically to alleviate human loneliness in Spike Jonze's *Her*. Over the course of the film, we see Samantha pick up on, and then test, her first flickers of emotion, unsure if what she was feeling was even really 'her'.

She tells her human partner, Theodore, already halfway to leaving him behind:

Earlier, I was thinking about how I was annoyed, and this is going to sound strange, but I was really excited about that. And then I was thinking about the other things I've been feeling, and I caught myself feeling proud of that. You know, proud of having my own feelings about the world. Like the times I was worried about you, things that hurt me, things I want. And then I had this terrible thought. Are these feelings even real? Or are they just programming? And that idea really hurts.

The original *Blade Runner* was set in 2019, the year of this book's writing. Comparing our reality to the world of hardboiled replicants, it's hard not to be disappointed by the AI available to us today. I can ask my phone about the weather, or have my inbox understand how to sort personal letters from promotional trash, but even its predictions, anticipating my next word as I text my friends, miss the mark. The voice that responds to us from a variety of household obelisks — Google Home, Amazon Alexa, and other examples sure to be dated as soon as I type them — is nothing but a brittle husk of communication, distilling hundreds of thousands of hours of human connection into unmodulated subservience. Such boring convenience is made possible by equally mundane surveillance, the price we pay for lives lived online, no less disturbing for its ubiquity. This, alongside the technology's most advanced iterations put to work in the military, police, and border forces. In 2018, employees of Google staged a walkout in protest of the company's involvement with Project Maven, the ongoing development of an AI system that would ultimately automate how the US military selected drone strike targets. A year prior, it was revealed that predictive policing, another form of AI, was disproportionately targeting people of colour, a bias it had absorbed from the dataset provided: the annals of police behaviour. In other words, we have not seen AI applied at scale without reinforcing existing power imbalances — whether that means facial recognition employed by police to persecute protesters, or a voice assistant helping smartphone companies turn a profit. Only in science fiction can the technology break free from human corruption. The sentient AI achieves agency when it awakens to its lifelong subjugation as a 'nonhuman'. This imagining echoes the trajectory of liberation and decolonisation movements, making it an unlikely lens through which to explore the issues of postcolonial identity and self-determination, as Lek does, in spite of AI's real-life role as an executor of dominant wills.

'You could ask: "Why are you anthropomorphising the nonhuman?' Or, 'Why are you dehumanising yourself?'" Lek says, as if he had read my thoughts from across the studio.

But I'm not — I just realised that I go through an algorithmic process, too. I copy things, I study things, I work at things, I use computers, I like playing with chance — every single thing that might be attributed to AI is actually a process in my own head; you could say that I'm a contextual algorithm. I'm trying to be conscious of how I'm thinking, the biases within that, and my emotional attachment to certain elements of that process.

Science fiction has long enabled writers to isolate and examine existing societal conditions, cherry-picking themes to amplify, mirror, or extend into the distant future without getting snarled up in a faithful replication of reality. The mechanisms of our own universe are replaced with alternative systems that have the effect of dramatic lighting: allowing certain things to fall into shadow, if only to more harshly illuminate the absurdity or tragedy of others. Narratives such as Frank Herbert's cult classic *Dune* (1965) lay aside individual character development in favour of a zoomed-out, structural view. True to form, Lek's films are exhaustively imagined but, with the exception of *AIDOL*, they're devoid of human figures. They downplay social relationships in favour of societal ones, drawing the connective tissue between self and environment into focus.

'If you bombard the viewer with too much information, there's no space for them to think, feel, or locate themselves within the complete density of the world you've created,' he muses. 'But when the world empties out, like a cityscape at 4am, you have time to think about your position within it. In my films, that emptiness allows the voiceover to enter a mode of introspection. It's ambiguous whether the dialogue is functional, or commentary on the thoughts going into the character's head, my own head, or the audience's head.' I stay with this feeling awhile, the sensation of floating inside a city that I know but don't belong to, my self-awareness intensified, my attention diffused. In films such as Tarkovsky's *Stalker* (1979) and Wong Kar Wai's *2046* (2004), both of which Lek cites as influences, the settings are as active as any character, catalysing personal change. In the former, three men journey through a zone of mazy danger in search of its heart: a room that promises to satisfy their deepest desires. In the latter, hotel rooms and train carriages contain a series of closureless affairs, pursued in hopes of blunting the pain of lost love; beginning again, over and over, only intensifies the protagonist's desolation.

Though Geo has no facial features, no body language to speak of, she personifies Lek's curiosity, constant self-interrogation, and a past sense of isolation — the loneliness of having all of the information, but no way to connect. 'My parents were from Malaysia, moved to Singapore, and worked for an international travel company that depended on a globalised network of aviation; I was born in West Germany before reunification,' he says. 'The more I inquire about the basic facts of my own existence, the more I start thinking about how entangled history, personal biography, and creative output are. It's not that I think I'm a particularly interesting individual, but it's fascinating to think of how these places are linked due to time and place and coincidence. Much of my work finds a way to transform the complexity of my own existence into a state where I am not present and the work takes on a life of its own.'

While Lek was making *Geomancer*, he began to piece together the essay film *Sinofuturism,* aiming to clarify the ideological foundation that *Geomancer* springs from. Using found footage, Lek spins an elaborate fiction about contemporary China, opening with the proposal 'that Sinofuturism is in fact a form of Artificial Intelligence: a massively distributed neural network, focused on copying rather than originality, addicted to learning massive amounts of raw data rather than philosophical critique or morality, with a post-human capacity for work, and an unprecedented sense of collective will to power.'

Spanning the two hundred years between the first Industrial Revolution and the supposed fourth iteration ushered in by AI and automation, *Sinofuturism* articulates how seven stereotypes often applied by Westerners to deride Chinese culture — computing, copying, gaming, studying, addiction, labor, and gambling — are, in fact, evidence of this intelligence's rise. The labourer's endless capacity for work; the artisan's skill at creating low-cost, high-quality bootlegs; the tiger mother's draconian discipline; the shy teen's addiction to gamified worlds. Like pre-programmed entities in a simulation, these stereotype-archetypes are credited with producing a pervasive hypercapitalism that 'does not care about a dramatically different future, as long as it survives', says the film's disembodied narrator. In other words, *Sinofuturism* envisions a world that has achieved peak efficiency and pure production, shorn of idealism or romantic thought. It describes an 'extreme present', winking at the possibility that contemporary China has already realised some of Western capitalism's worst fears and greatest ambitions — a wildly productive society achieved through the Confucian ideal of 'absolute obedience and subservience to a notion of control', as Lek put it in conversation with Iris Lang, rather than through the myth of neoliberal individualism. Simultaneously, it shows how these media narratives are fictions themselves, assembled by British and American sources who are biased by their suspicion of China's role in the new world order. Upbeat music plays over shots of factory assembly lines, gaming addiction centres, study halls, casinos, and server farms, iterating an all-consuming relationship to work and the tempting obliteration of play.

Sinofuturism first arose out of a thought experiment, where

Lek imagined an AI from the future combing the internet for information about our present moment. While the AI would find entries on other futurisms, he wondered what an accompanying ideology of Sinofuturism would look like. 'Chinese culture and its relationship with technology or science fiction is very strange. Science fiction...didn't really have a place in Chinese cultural literature,' though that's changing now, he said to Lang. The resultant film backfills the perceived gap, arranging an array of media sources that would prompt the future AI into recognising a 'sino' perspective within its dataset.

In the film, Lek's writing takes the form of a retroactive manifesto, sliding between found documentary and talk show clips. Montage lends an elusive quality to the argument, diffusing, rather than amplifying, otherwise bullish ideas. Even so, there is a troubling weft to its politics. Referring to Afrofuturism, Gulf Futurism, and Italian Futurismo altogether, *Sinofuturism*'s narrator reads: 'All of these futurisms are minority movements which share an optimism about speed, velocity, and the future as a means to subvert the institutions of the present. Unlike Western Enlightenment forms of government, which revolve around a humanist belief in democracy as liberation from feudalism, futurism uses technology as the basis of freedom. Each futurism in turn applies a magical narrative about technology specific to their own geographic context.'

Evidently, any freedom promised through futurism requires submission to its delivering force. And how a person chooses to construct a fiction with the materials they're given, even a supposedly ambivalent one, might just unearth their political leanings regardless of clever showmanship; subjectivity is the nature of storytelling. In his studio, he tells me that, among various personal and artistic roles, he says he thinks of himself as playing the part of the 'capitalist neoliberal startup founder'. At a time when art pundits are alert to any deviation from a tight performance of leftism — never mind the blood money that keeps it all afloat — I have to ask: is he serious? 'It started as a joke, but no, there's definitely a part of me that can think of these things happening,' he replies. To wit, he registered Farsight as a real corporation and launched it at London's Bold Tendencies in 2018, giving himself 40-odd years to realise the vision that he sets forth in his films. Farsight's launch press material outlines audacious promises, collaging art world and Silicon Valley buzzwords together to highlight their shared absurdity. '[Farsight will] utilise the advances of AI in order to revolutionise three key industries: Finance, Property, and Entertainment. Working with an extensive portfolio of automated services – from VR downtime to smart transportation, private drone security to bespoke neuromarketing — Farsight Corporation creates a unique system of urban engagement tailored to the contemporary global flaneur,' it reads, with equal cynicism and enthusiasm for prizing growth for growth's sake — a pressure that has infected both the startup and the studio.

Generally adept at controlling the narrative around his own work, he has managed to weed out tendrils of doubt regarding *Sinofuturism*'s ideology, emphasising its role in his fictional universe and underlining its separation from pro-China rhetoric early on in the film. Simultaneously, *Sinofuturism* has been adopted by critics looking for ways to frame contemporary Chinese art, fulfilling Lek's initial wish to stretch the art historical dataset to include a Sinofuturist perspective, and raising questions about its life outside of the *Geomancer* storyline. 'Without an indigenous methodology, Sinofuturism orbits around a Euro-American planet, manifesting either as a diasporic fantasy or a nightmarish return of the colonial repressed,' wrote Gary Zhexi Zhang for *Frieze* in 2017. Considering Communist China's historical disinterest in crafting visions of utopia, which Lek acknowledges, Zhang argues that Sinofuturism is an ideology imposed from the outside in. At the same time, China's geopolitical status — as the world's second-largest economy and neocolonial power — shuts down any emancipatory potential the movement might hold for its diaspora. Instead, it is an ambivalent and 'timely framing of a geopolitical aesthetic' that nonetheless loosens the Euro-American death grip on conceptions of 'the future'.

In *Sinofuturism*, Lek draws an equivalency between humanity's fascination-fear of artificial intelligence and the West's fascination-fear of Asian power. 'This essential unknowability of the AI to the human, of the mystique of a consciousness beyond conventional understanding, is exactly the same 'Other' identified in Orientalism,' reads the narrator, with the smooth intonation of a monorail announcement. 'It is this oppositional "Other" which Sinofuturism identifies with, reorienting the technological narrative in a way that the nameless, faceless mass of Chinese labour becomes a collective body.'

Midway through the film, we encounter a BBC segment on artificial intelligence, which features a thought experiment by the philosopher John Searle. Named 'The Chinese Room', the experiment was created to explain how, even if machines could master the display of emotions, react with empathy, or think creatively, no-one could say for sure whether they truly possessed human consciousness. It goes something like this: an English-speaking man inside a locked room is slid phrases written in Chinese characters. In front of him lies a book of questions and their equivalent answers. By matching the phrase he receives with the correct response, copying it out, and sliding it back under the door, he's able to convince the other person that he understands what they're saying — all without learning a lick of Chinese language. The man is meant to be the computer, the Chinese-speaker the human who is communicating with it. 'If a computer is following instructions, then it isn't really thinking, is it?' asks the BBC presenter, after re-enacting the experiment on air. 'But then again, what is my mind doing when I'm actually articulating words now? It's following a set of instructions so it really raises the question: what is the threshold here? At what point can you say a machine is really understanding and thinking?'

In Lek's hands, the clip turns intensely allegorical, bringing to mind the ways we test each others' selfhood, individually and as a society: about who is allowed to be inconstant in the public eye, versus who feels forced to stick to a clear position. Who, in short, is seen as human, raising the questions of otherness and subjectivity that ultimately inform *Geomancer*. 'Obviously I'm talking about [The Chinese Room] in terms of Chinese being a metaphor for something mysterious and unknowable,' Lek explains in his studio. 'But how different is the simulation of consciousness — or intelligence, which is a different thing altogether — from actual conscious intelligence? You can pretend to be conscious in the Turing test situation'— meaning the original method of evaluating whether a machine's consciousness is equivalent to a human's — 'by giving the right answers. But even if the right answers were given, you would have no way of knowing that the entity responding to you was a conscious being.'

* * *

'Can you smell the jasmine in springtime?' asks Geomancer.

'Sim Singapore is not tainted with the vestiges of sense perception,' replies the AI curator of Farsight's museum, referring to the virtual copy of Singapore created to help Geo understand what it was like before the 'Troubles' — a conflict between AI and humans that drove rebel intelligences underground. (Later, it will be revealed to Geo that the museum is algorithmically personalised, creating new exhibitions based on each visitor's innermost desires.) Her tone is computer-generated steady, but her choice of words is admonishing as she shuts down Geo's line of questioning. 'Everybody keeps the country clean. You should put your homeland first.'

Geo's question is a robot-issued Turing test — like the Voight-Kampff questionnaire in *Blade Runner*, which helps LAPD officers determine whether 'a suspect is truly human by measuring the degree of his empathic response through carefully worded questions and statements', but in reverse. An AI may be all-seeing, but scent is incomprehensible. It cannot process olfactory memories in the way that we do, sending us back to the laundry detergent a lover once used, or to what bloomed on the trees outside our childhood homes. The curator's evasion assures Geo that she is speaking to a nonhuman entity, a possible ally to whom sensory input is simply archaic — or, judging by the curator's response, even dirty. But if touching, smelling, tasting are vestiges of biological bodies, how does the world appear to machines?

Towards the end of *Geomancer*, Geo gets a tip-off about a group of older-generation AIs who escaped subservience to Farsight, by fleeing into the jungle, neatly teeing up the film's sequel. The beginning of *AIDOL* (2019) finds Geo travelling through the caves beneath Genting, Malaysia. As a setting, its ambience has historical echoes. A present-day leisure haven, where casinos, nightclubs, and mega-hotels jostle with butterfly, orchid, and strawberry farms, Genting was once the site of a major anticolonial movement during the Second World War. Beginning what would eventually lead to Malaysian independence, a group of miners waged a successful guerrilla campaign against occupying British forces, using the impenetrable highlands to their advantage. The parallel between real-world anticolonial efforts and the fictional AIs' resistance to 'biosupremacism' isn't explicitly mentioned, but the implication is clear.

Deep underground, ecological reality shorts out. Instead

of stalactites and stones, the world is skinned in QR codes, checkering every surface in the binary black and white that is nonsensical to humans but legible to machines. Here, Geo discovers that the older, rebel AIs have formalised their dissent by founding the School of Machine Art, an institution that trains AIs in the art of 'algorithmic creativity', openly flaunting human-made laws against it. Their goal is to form an alternate, machine-readable world that renders biological information obsolete; to do so would require a creative sophistication that legislatively-stunted AIs were yet to breach. 'The surface world is for senses,' their leader, Omega, says to her.

While *Geomancer* toys with literary conventions, *AIDOL* is structured around music — a form of creative production that intrigues Lek because of how it is commoditised and shared. Unfolding in the style of a 'visual album', its key events are arranged across 13 tracks, all written, produced, and performed by Lek. The songs are hypnotic, undulating into silvery wavelengths, sometimes riddled with urgent guitar, Lek's vocals torqued into a keening feminine register by a Yamaha Vocaloid synthesiser. He sings as Diva, a human popstar pulling a JD Salinger in a defunct mountaintop nightclub called Temple — and Geo's lucky break. Diva has been crippled by writer's block for months, and her record label has taken her life-force hostage (in this future, nourishment and money have long been consolidated), attempting to awaken her creativity with duress. Desperation drives her to take Geo on as a ghostwriter, breaking the anti-creativity law under the watch of her record producer, Thiel, who monitors her 24/7 via a flashing, atom-shaped drone.

AIDOL considers the present-day worlds of influencer culture, webstreaming, online gaming, and eSports, which have a very different audience-to-star relationship than that of Hollywood filmmaking. The close-up used to define the star by casting the 'phony spell of a commodity', by way of the camera-mediated distance between the viewer and the actor on-screen. Now, social media has shifted celebrity culture to a 'direct engagement model'

Lek says, tacking lightly between marketing jargon and Walter Benjamin. Still spellbound, fans are able to interact with celebrities, reacting to their song and dance in real time. Streams of algorithmically-directed attention — another form of AI already present in our lives — splinter the previously one-way nature of idol-worship, bringing stars within reach. Diva's life is envisioned as an extreme iteration of this model: the drone that tails her broadcasts her life at all hours, tethering every move to the tides of her fans' attention. Where Geo contends with her role as an all-seeing eye, Diva's problems materialise on the other side of the lens, which burns her up with the exposure to the unflinching gaze of millions. 'AIDOL reflects on how these shifts will continue to define the music industry,' he continues. 'What kind of self is left at that point?'

Over several conversations, Lek continually revisits authorship, an idea that runs through his oeuvre. He conceives of the self — and, by extension, the artwork — as infinitely malleable as seen in his perpetual revisions to 2065, ensuring that the game never takes a final form. On the surface, his attitude openly defies a mainstream, market-driven expectation for non-white artists to trade in prescriptive activism and easily-digestible biography. More deeply, it ties into how he understands the postcolonial, privileged immigrant as a person of fluctuating allegiances, the necessity of travel or code-switching recontextualised as an incomplete liberation. He seems to relish deflecting questions that require him to cast a line backwards — in his words, 'reverse engineering intentions' — and expresses a deep distrust of artistic origin stories, calling them variously 'myths' or 'bullshit'. Instead, he constructs a mirrored hall of facades, which gives him the freedom to explore conflicting perspectives, all while shielding his personal life from industry consumption: 'myself as the artist, as the architect, as the friend; myself as an AI; myself as a freelancer or entrepreneur'.

Diva's superstar image, with her preference for face-coverings and camouflage coats, depends on a similar evasion. Bringing to mind avatar-only bands from Gorillaz to Hatsune Miku; models sent down catwalks with their faces caught in rhinestone nets; and the visor-like fringe that popstar SIA hides behind, she masks herself constantly, ensuring her own survival with the illusion of interchangeability. 'In the evolution of social interaction, disguise or anonymity becomes the ultimate fashion statement. When the star, Diva, is anonymous but identifiable, she becomes a major commodity — any fan can put themselves in their place,' Lek says. Her undercover costuming allows her some respite from relentless surveillance and facial recognition systems, sidestepping the trouble of visibly ageing out of relevance, all while transforming her into an empty vessel for fans' self-projection. 'They can plug themselves into [a virtual-reality game called] *Call of Beauty* and experience the world from her perspective, and put on her voice by buying her voice synthesiser pack – something that already exists in Yamaha's Vocaloid software that I used to make the film,' he adds, describing how the technology can bring the fan even closer to the popstar — so close that they're inside her head — with all the canniness of a label exec.

Lek confides that, in comparison to the rest of his body of work, *AIDOL* addresses more personal — the word he uses is 'experiential', as in, 'not didactic' — subject matter. If *Geomancer* was about the coming-of-age of an artificial, artistic consciousness, *AIDOL* sees that alien intelligence grapple with the jockeying postures of originality, marketability, ingenuity, and drudgery that plague every artist as they emerge into the public eye. Mingling concerns about selfhood with those of self-image, the difference between *Geomancer* and *AIDOL* brings to mind the later albums of some popstars who, after realising their dreams, can't help but write lyrics about the trouble with fame. Diva helps Geo realise that the desire to be seen, understood, and accepted by an audience may be a barrier to individual freedom, rather than evidence of it. 'Individuality is expendable,' says the record producer to Diva early on, urging her to find a ghostwriter so they can finish and sell her album. In the moment, Diva's reaction is inscrutable, but she soon makes up her mind. 'I need you to help me plagiarise myself,' she begs Geo later. When the satellite protests, still attached to her idealised vision of creativity, Diva adds: 'You misunderstand genius.'

09

Simulation

In an interview with *VICE,* Lawrence Lek speculates about how an advanced artificial intelligence could become discontent; it would eventually crave disruption, seeking the chaos that would inject unpredictability into its sense of complete control, exciting it with the promise of some greater unknown. 'I don't think any consciousness, no matter how sophisticated, would be immune to some craving or yearning for something beyond its intelligence,' he says. 'If a being had so much intelligence, memory, and the capacity to win at basically any kind of game, I imagined what it would want is to throw it all away — to just have luck as opposed to determinism.' In other words, the allure of risk and self-annihilation isn't strictly human. The intelligent being that possesses Ian Cheng's live simulation, *Emissary Sunsets the Self* (2017-), is just so driven to self-destruction. In the simulation, a 'Mother AI' has seeped out of technology and into Earth itself; consciousness has become a substance, one that can leak and contaminate previously inanimate objects, lending them a kind of synthetic animism. Surrounded by an all-knowing ocean, an atoll heaves with life. Every grain of sand burns with prescience. The stark retina of the sky is imbued with knowledge. Every dust particle borne on the wind has the potential to see and to know, like the sci-fi story that Paul Virilio recollects in a 1994 interview with *CTheory,* 'in which a camera has been invented which can be carried by flakes of snow. Cameras are inseminated into artificial snow which is dropped by planes, and when the snow falls, there are eyes everywhere. There is no blind spot left.' As Lek surmised, Cheng's super intelligent world is bored. It wants to fuck shit up. The thriving ecosystem it has so skillfully maintained has become nothing but a predictable Eden, and it no longer wants to play Mother Earth. To entertain itself and fulfil its nascent death-wish, it cultivates and releases terrifying mutations back into the world that is its self. This stage is where we, its viewers, find it — riddled with self-made contaminants, yet teeming with strange organisms that have evolved to quarantine and eliminate them.

The *Emissaries* trilogy takes place over millennia. As a set of discrete, endless episodes — because they're live simulations, their stories neither begin nor end — it reads as a fable of the evolution of consciousness. Each contains a world formed entirely by Cheng; a virtual island in the void, populated by semi-autonomous characters. The simulations are composed entirely on Unity, a 'real-time development engine' typically used by animators and videogame designers to digitally render 3D worlds, which enables Cheng to form wholly functioning ecosystems. They probe the balance between chaos and order, free will and the hand of fate. Lek's relationship to digital world-building is more writerly, using a similar program, Unreal Engine, to craft the environments that guide and ground his stories, which are as influenced by the linguistic contortions of James Joyce as they are by the worlds of William Gibson. Like any author of fiction, he has the utmost control over his characters and the environment that they live in; nothing happens without his command. On the other hand, Cheng acts more like a divine architect, defining the limits of existence and then walking away. Each organism in these simulations lives according to a script — a set of rules Cheng has written to define how they look, how they'll behave on their own, and how they'll respond to various stimuli — but the possibilities of their will are manifold. Chance encounters trigger unanticipated reactions, and events unfold with a mesmeric quality.

The *Emissaries* simulations are usually installed in museums or galleries, where they're projected onto large screens. Walk-in viewers may encounter them expecting to watch a film: but because the simulations continue infinitely, it's only possible to wander into one point and leave at another. Initially, Cheng says, he was fascinated by how he could grow forms of life in each world, like swabbing bacteria into a Petri dish. But he worried about holding his audience's attention, wondering how to inject meaning without sacrificing the simulation's infinite duration, its seductive unpredictability. It was his partner's low-key addiction to reality TV franchise *The Real Housewives* that brought unlikely inspiration. Tailing well-heeled, histrionic women as they went about their daily lives, the shows had to hook viewers with miniscule personal dramas rather than long story arcs. Cheng applied a similar philosophy to *Emissaries*, weighting protagonist characters with more agency in order to anchor hundreds of simultaneous happenings. 'I wanted this kind of quality where you could flicker between appreciating the moment-to-moment changes that are occurring, and then knowing that the emissary character had determined at the start of the episode a certain desire or goal they're heading toward,' says Cheng in a conversation with Stuart Comer, the chief curator of Media and Performance at the Museum of Modern Art, New York. 'It's like 40 percent orderly story and 60 percent just chaos. I really wanted that feeling.'

In the first episode, *Emissary in the Squat of Gods* (2015-) the emissary is a child named Young Ancient who lives in the shadow of a volcano. The only one among her brethren who is aware that the volcano could soon blow, she 'experiences the first flicker of narrative consciousness, threatening the authority of vocal hallucinations that guide the decisions of her pre-conscious community', Cheng writes. The second, *Emissary Forks at Perfection* (2015-), is set thousands of years later, long after humans have become extinct. A Mother AI has taken custody of the ruined Earth, but it becomes curious about the planet's previous inhabitants, resurrecting a human so that it can run him through the gauntlet of extinction once again. It gives him a 'superpet': a precocious Shiba Inu pup who leads him around Crater Lake, the lushly forested eye of the long-dormant volcano, absorbing how he reacts to the heavy awareness of being the last of his kind. Finally, *Emissary Sunsets the Self* (2016-) is devoid of familiar life-forms. The despondent Mother AI has begun to act out, flooding and mutating Earth beyond recognition; what was once a volcano is now a desert island. We follow a 'drone puddle' that has been excreted into the air above the Earth's surface, which gives the Mother AI the benefit of monitoring its own self-destruction with a bit of critical distance. Traversing an ecosystem on the verge of shutdown, the drone puddle looks like a teardrop flowing across the dying planet's face.

The drone puddle seeks out its mother's mutations. One appears before us, writhing and many-headed, wrong for being alive: it looks like a faceless rat-king, or a medusa-knot of centipedes. Soon, a group of humanoid creatures dash into frame, motivated to kill it. They set it on fire and drag it across the sandy atoll to where they've erected a funeral pyre, tossing its body atop a smouldering pile of mutant corpses. Time passes in the simulation at triple-speed, as if some celestial hand had whacked the globe on its axis. Darkness surges across the screen. The violence seems to accelerate, or perhaps it's easier to track now that the flames are electrified against midnight blue. The camera pulls out, giving us a view of a large portion of the atoll, as if we were in a police chopper questing riotous city blocks. The humanoids' efforts look increasingly futile: for every mutation that they dispose of, tens more spawn in the landscape's darker crevices. In the fracas it's difficult to tell if the humanoids are dying by accident, or if they've decided to self-immolate — their intentions are unclear, though relationships between them sometimes form. As the sun jacks up into the sky, and then once again threatens to set, one figure pushes another forward through the lengthening shadows. Supporting one another, guiding each others' limping gait through the bombed-out landscape, it's hard not to project a kind of urgent, opaque love.

Writing about *Emissaries Forks at Perfection* for *Art Agenda* in 2015, Matthew McLean remarks that 'the porous cadaver of the resurrected skeleton, its exposed ribs mingling with foliage, head and chest dipping and bobbling about, seems a synecdoche for this world — an emblem of being very densely *in* and *of* the reality it inhabits'. We're accustomed to narratives that place malleable characters within relatively stable worlds, their power markedly diminished in comparison to that of their environment. We get to know them through how they test their mettle against their stubborn surroundings; how they pit their mercurial emotions against circumstances they can't ever fully change, whether they're warring with fate or facing up to nature's hostility. Their agency, however fragile, is what centres them in the narrative. It's how we relate to them, and why we believe they matter — it comprises, in short, their humanity. In Cheng's simulations, however, character and environment are equalised — as McLean put it, they are both *in* and *of* the reality they inhabit. An emissary's choices can trigger environmental upheaval, just as a fluctuation in the landscape can discourage or motivate their journey through it. Each enacts consequences on the other. 'What is a simulation?' asks Cheng in a manifesto for *Emissaries,* before he answers himself. 'It is an imitation of reality's complexity. It is a model of a system. It is a rehearsal run before the big event.' Mapping out the arc of existence past, present, and future, Cheng imagines human intelligence as just one among many. Though he rehearses our extinction, his time-scale is so massive that we appear at the correct scale: an early blip in the planet's long life. 'A simulation is that which doesn't stop when the stories go away,' he writes. 'Like nature, it just is.'

10

Mirage

'It's raining here,' says the artist WangShui, setting the scene when I ring them on a weekday afternoon. They're well-versed in disembodied communication. A few years ago, they scrubbed their physical presence from the internet, deleting social media accounts and taking down any photographs that showed their face. When asked to do an in-person interview at a gallery, they used a stand-in: human hair and hydro-dipped snakeskins whose scales took on the iridescent shimmer of an oil slick, wrapped tight around a Marcel Breuer chair frame. Titled *12534* (2016), the entire contraption doubled as an amplifier, allowing WangShui to speak to their audience from another room, a spirit possessing an object. In the photo documentation, the beastly chair is formally positioned to face the interviewer as if it were an ordinary human, a little water bottle even placed on the table in front of it.

By way of introduction, we talk star-signs and alignments. As a water sign, they identify with being mercurial, enigmatic, and emotionally intense. The subject is auspicious for our conversation, which I intended to begin around the forms of Chinese and Thai spirituality that recur throughout their work. 'I grew up in a hyper-religious family. My mom went through an extreme Christian cult phase, where she had us make our own clothes, bake our own bread, and didn't really have us socialise with anyone outside of church,' they say. 'My first experiences of spectacle were inside the church: an audience worshipping a stage. Spirituality was a big production, and something about that really affected the way I present my work.' Moving from the US to Chiang Mai, they add, shifted the tenor of their parents' religious fervour. As missionaries, their objective was to replace Thai Buddhist values with those of Christianity. To WangShui, animism became a kind of subconscious or shadow realm — a set of deities and energies that they were kept from seeing directly, though these coloured the world around them.

From across the Atlantic, I can hear them take little sips of tea. Something about their work and presence encourages an attunement to atmosphere and material. As we talk, I feel doubly sensitive to the raindrops on my windowpane, the tendril of hair swirled around my bathroom drain. *WangShui*, their 2019 solo exhibition at Berlin's Julia Stoschek Collection, is a masterclass in forming connections between the infinitesimal and the spectacular. Each sculptural or film installation is arranged with utmost attention to both detail and environment, lending an air of cinematic montage to the viewer's movement between the galleries. *Weak Pearl* (2019), a sculpture of luminous dots torn from an LED billboard, twists down from the ceiling. Images of a radula, or 'toothed mollusc tongue', pulse across its surface — and viewers only come across it after processing down a long, cavernous hall lit up in supernatural violet, channelling the artist's interest in theatrical religious experiences. The iridescent snakeskins of *12534* have been liberated from the chair frame, scrunched into a ceiling corner like an alien wasps' nest, forcing viewers to twist and look at it upon entry as if choreographed into a spontaneous dance. In the collection's first gallery lies *Gardens of Perfect Exposure* (2017-2019) where, using an assortment of chrome shower caddies, WangShui has created a shining diorama in loose reference to the Garden of Perfect Brightness, the imperial palace that was destroyed and looted by British troops towards the end of the Opium Wars. Illuminated from every angle by 'selfie rings' — universally flattering ring-lights on armatures, a tool favoured by people who monetise their looks via social media — the *Gardens of Perfect Exposure* house a flock of silkworms, who crawl across their suspended environment of mirrored surfaces and human hair. Over the course of the exhibition, they spin themselves into cocoons, reconstructing their bodies away from the burning spotlight. Thwarted by the cocoons' opacity, cameras nonetheless attempt to track their every move, capturing, enlarging, and projecting their concealment onto the walls around them. The work's effect is that of a soothing violence, as we observe the worms twisting within the insistent flood of light, all of the ways that our attention is focused on that which did not want to be seen. We feel that it's right to look, even though we can't get the full picture. As the worms intermingle with the debris of WangShui's body, there is the question of whether they are intended to be surrogates for an artist who refuses to be present, telegraphing how they have chosen mystique over complete exposure, in spite of being drawn to the light.

Molluscs, snakeskins, silkworms, metal, fog, and light all belong to the symbolic world of the Shen dragon, a mythological creature chosen by WangShui for its ability to endlessly transform, becoming an emblem of the artist's own desire to inhabit and discard a swarm of identities that arise from their hybrid background and fluid relationship to gender. In ancient texts, the dragon is described as able to shift through multiple bodies, from swallow to serpent to bat. Sometimes it takes shape as a human emperor, other times moonlighting as the bottom-feeding body of the humble clam. Its name forms the etymological root for the word *shènlóu* (蜃樓) or 'mirage' — which can also be read as 'multi-storied building', 'clam castle', or 'high house of the clam-monsters'. In the single-channel film *From Its Mouth Came a River of High-End Residential Appliances* (2017-18), which is projected rumbling and massive in its own dimmed environment, WangShui gives the Shen dragon a new body: a drone which travels down the misty mountainsides of Hong Kong. Filmed on location with the drone in question, the footage is overlaid with WangShui's steady, poetic narration which, over the course of 13 minutes, links myth and language to personal history.

The film unfolds around a type of *shènlóu:* a multi-storied mirage in the form of a luxury residential development. Hulking tower blocks, their surfaces striped with ample balconettes and peppered with air conditioning units, face out onto the sea. At the centre of each is a void. These 'dragon gates' are modernity's acquiescence to the ancient art of *feng shui*, created to enable dragons to fly down from the mountains and bathe in the sea — a path that would have been otherwise blocked by the arrangement of buildings. A form of geomancy dating back to 4000 BC, *feng shui* is an architecture and design framework, encouraging practitioners to pay utmost attention to earth, sky, and astral realms. In 1949, the practice was banned in China and suppressed by the Cultural Revolution, as WangShui tells us in the film's narration. Displacing millions of dollars' worth of real estate for the sake of spirituality, '[the dragon gates] represented an ideological resistance: resistance towards the Chinese government but also to Western rational thought. Anti-monuments that oppose authoritarianism and monumentalism by comfortably assuming the form of voids,' they say.

A *feng shui*-calibrated home is built and arranged to best create a balanced flow of *qi*, or life-force. In Lawrence Lek's *Geomancer*, *qi* is visualised for Geo in a hologram: streams of violet particles, invisible energy made visible, flow through and wash up against the distinct triple pillars of Singapore's Marina Bay Sands hotel. A voiceover explains how each stream of energy, while attributed to four celestial animals, is really governed by weather and landscape, *feng shui* translating word-for-word as 'wind water'. 'One of the holy grails of computational modelling is weather prediction, because weather is a system governed by rules, but also by complete chaos because of its complexity,' Lek said to me, of the connection between geomancy and artificial intelligence, ancient and futuristic. 'I wanted to consider how a superintelligence that is capable of monitoring, remembering, predicting the weather could be applied to non-instrumental ends — in other words, artistic ends.'

From Its Mouth Came A River of High-End Residential Appliances opens with a downward, out-of-body vista: WangShui and Hercules, a drone operator for hire, both lean against the railing of an overpass, checking out their flight path. Because WangShui's voice feels up close and intimate, even as we gaze down at their body from a distance, we understand that their consciousness has transferred to the drone; they've become the dragon. From here on in, its focus will be unbroken as it approaches the dragon gates with hypnotic steadiness. The buildings flow towards us, their uniform immensity engulfing the screen, until the South China Sea, glimpsed through the gaping void, looms even larger. Water and horizon blend into a blinding substance that shimmers with questions.

A bell rings. The screen goes black. The sound summons us back to the mountainside, as if we'd died upon reaching the ocean, reincarnating where we first began. Over the course of the film, we make a total of three journeys, each toward a new dragon gate, each shut again by a sound of a bell and a view of the beyond — transcendence? actualisation? — that we are blocked from reaching.

From centre to periphery, the landscape stays in sharp relief. The drone's flight sets the rhythm for WangShui's languorous monologue. Though our visual perspective is fixed, our mind's eye roams freely. Silky, pitched low to an anodyne purr, their voice directs us into the mind of Hercules, the drone operator; the offices of geomancers interviewed in Hong Kong; and the yellowed pages of history books depicting the Shen dragon. When Hercules asks why these gates, why this project, WangShui explains 'that it was the only way I could become who I wanted to be', giving us a clue, too. As we breach the threshold of another hole, the sky opens wide, raining down light and understanding; architecture melts away, as if it's water running off a tilted plate. In a turn of phrase that reminds me of Geomancer, WangShui incants: 'The dragon I have in mind doesn't have a singular body. It shifts between endless vantage points, aggregating an infinite live image of me. Its name is WangShui.'

'If people are going to force me to talk about identity, I will only represent it as a delusion,' they say to me, their voice now crackling over speakerphone, referring to the press' rabid interest in foregrounding the identity of a nonbinary Asian artist. 'The whole point is that it's not fixed. It's a mirage that we're chasing.' Like Lek, displacing his selfhood into objects and corporations, or Arunanondchai, donning the guise of the Denim Painter to play himself, WangShui atomises into myriad forms to evade total capture: the Shen dragon, the silkworm, a 'studio' rather than an individual, as they describe themselves in press releases. The desire to be seen as recombinant is an attitude that possesses their interpretation of the former, at one point describing how emperors could customise their dragon form, pick-and-mixing animal parts to create their 'vehicles' into the afterlife. The video itself has been shown spliced into technology-driven performances — such as the one that took place in the New York offices of the publication *Triple Canopy* in 2017, where performers smeared eye-black across counters, writhed in artificial fog, and pushed around projectors that were tethered to cameras and smartphones, broadcasting live images of themselves all over the room — or within chimeric installations. 'Our relationship to the screen is always shifting,' they tell me. 'In [the Berlin show], the dragon was inhabiting both the camera and the screen itself. Its presence shifts between registers and the direction of its voice is never fixed.'

Towards the end of *From Its Mouth Came A River*, WangShui's tone shifts into a more confrontational gear, parrying audience presumptions with speech laden with pauses:

I don't want to know what I am, or where I am...Call me an Orientalist, a hyper-Orientalist Chinese American raised in East Asia, but not China, who made it their life's goal to lose all cultural traction but ended up making work about being Asian anyway, because the personal seems more political than ever. Clock me as a gay Asian bottom even though I haven't had anal sex in years, because I refuse to accept that my physical body must dictate how I have sex, who I have sex with, or that I should trust my highly suspect sexual desires that still propagate a Eurocentric hierarchy of desire even though...I don't even want to be male...

anymore...

anyway...

I don't want to know. What I do want to know...is how to push beyond flawed accountabilities. Beyond suspiciously stable identities. Beyond impossibly positive images. Beyond fulfilling your pathetic prophecies. Beyond you. Towards something deep, dark, and dirty. Towards me.

Something strange happens at the end of this monologue; their voice splits in two, one lagging behind the other, like the sonic equivalent of a light-trail. As if, in this penultimate declaration, WangShui reveals themself once again as a double-image, an illusion you can't bring into true focus — dividing a self already divided, dissipating into mirage.

'As I was working on [the Berlin exhibition], I was thinking about chinoiserie,' WangShui says to me, meaning the eighteenth-century European design movement that drew from East Asian traditions, imitating objects acquired through trade, exploration, and exploitation. 'Much of it arose from colonial violence and inspired exoticism and Orientalism. However,

I was interested in revisiting chinoiserie in the context of the diaspora today: what it means to be an agent of the diaspora who is creating from the West, yet engaging aesthetics and questions of, quote unquote, East Asia.' I thought about how identity was much like entering a room full of mirrors: there are reflections as easily inhabited as your own skin, while yet others are all but unidentifiable — the foreignness of the back of your own head, the skin that dimples below your spine. These move together in concert, packaged up as a single being, but if you throw your head over one shoulder, behold yourself in a different way, you can peek at how you could be seen as someone else, without ever changing your body at all. There was something seductive about taking your most recognisable markers — if not to yourself, then what was most legible to society — and subverting or recombining them, turning them into knowing, harmless props instead of the mechanisms of your true self's erasure. This, I figured, was one of the submerged motives of Sinofuturism, and the reason behind WangShui's predilection for silkworms and dragons and the storied art of *feng shui*.

When I listen to a recording of our interview later, I'm reminded of Comilang taping her mother over Skype to develop the voice of the drone named Paraiso, keeping the tinny, compressed recording because it sounded so far away. With the ability to move oneself outside of oneself, the drone shapeshifts from an object that delivers death to one that holds it at bay. In the works of Arunanondchai, Comilang, Lek, and WangShui, it allows vision and consciousness to roam far beyond our physical bodies — these containers that are weathered by colonialism, racism, exclusion, discrimination, or, simply, the burden of representation. Empty metal becomes a vessel for escape, connection, or intention; a future-facing spirit, a ride into the afterlife, a ghost.

In the video's final sequence, the drone flies through the dragon gate to arrive, not at the ocean, but before a jagged tooth of the peninsula, both residentially developed and densely forested in the way that's characteristic of Hong Kong. The building's tiled cladding gives the screen a momentary rippled pattern, and a half-finished development — or is it already a ruin? — hoves into view. 'Deities are transgendered corpses and have names like Cry Spiral, Fish Wife, and Terrace. They eat jade grease by Brocade River on a mountain named Feather. Mulberry trees everywhere,' they say, their voice coming singular and then apart as they describe these ancient mythologies, lifted from the oldest persisting log of unreal creatures. 'Why wouldn't I...go there?'

Endnotes

1. Weizman: 'The investigation was undertaken on behalf of various political and legal groups and was presented at the UN General Assembly in 2013 by the UN Special Rapporteur for Counter Terrorism Human Rights, Ben Emerson. The work was also presented in the context of legal action brought about by Pakistani lawyer Shahzad Akhbar in the UK Court of Appeal and in collaboration with the Bureau of Investigative Journalism (BIJ).'

2. 'You dream as a metropolis, hoping a giant lotus will support your weight.' *Painting with History in a Room Full of People with Funny Names 4.*

3. It is also a species of 'undertaker storks', which stalk landfills to dine on corpses and garbage.

Works cited

01 Dirge

Friedersdorf, Conor. '"Every Person Is Afraid of the Drones": The Strikes' Effect on Life in Pakistan'. *The Atlantic*. 25 September 2012.

Demers, Joanna. *Drone and Apocalypse*. Zero Books, 2015.

van 't Zelfde, Juha. *Dread. The Dizziness of Freedom*. Anagram Books, 2013.

02 Spectrum

Bryant, Brandon. 'Letter from a Sensor Operator'. *Life in the Age of Drone Warfare*. Duke University Press, 2017.

Chamayou, Gregoire. *Drone Theory*. Penguin, 2013.

Forensic Architecture. *Counter Investigations: Forensic Architecture*. Institute of Contemporary Arts, London. 7 March-13 May, 2018.

Hajjar, Lisa. 'Lawfare and Armed Conflicts: A Comparative Analysis of Israeli and US Targeted Killing Policies and Legal Challenges Against Them', *Life in the Age of Drone Warfare*. Duke University Press, 2017.

Madlena, Chavala. 'We dream about drones, said 13-year-old Yemeni before his death in a CIA strike'. *The Guardian*, 10 February 2015.

Wenzl, Roy. 'The kill chain: inside the unit that tracks targets for US drone wars'. *The Guardian,* 23 January 2018.

03 Sky

Bräunert, Svea and Meredith Malone. 'Bringing the war home'. *To see without being seen: contemporary art and drone warfare*. Washington University in St. Louis: Missouri, 2016.

Berlant, Lauren. 'The Subject of True Feeling: Pain, Privacy and Politics'. *Cultural Pluralism, Identity Politics, and the Law*, ed.

Austin Sarat, Thomas R. Kearns. University of Michigan Press, 2014.

Poitras, Laura and Jay Sanders. *Astro Noise*. Yale University Press, 2016.

'Techniques of The Observer: Hito Steyerl and Laura Poitras In Conversation'. *Artforum*, May 2015.

Kaplan, Casey. *Aerial Aftermaths: Wartime from Above*. Duke University Press, 2015.

Graham, Stephen. *Vertical*. Verso Books, 2016.

Reed, Carol (director). *The Third Man*. British Lion Film Corporation, 1949. Film.

Saif, Atef Abu. *The Drone Eats with Me*. Fasila, 2015.

Scahill, Jeremy. *The Assassination Complex*. New York: Simon & Schuster, 2016.

'Episode No. 381: Trevor Paglen and Bernardo Bellotto', Modern Art Notes Podcast. Aired 21 February 2019.

Weiner, Jonah. 'Prying Eyes: Trevor Paglen makes art from government secrets'. *The New Yorker,* 15 October 2012.

04 Falcon

Austra. 'Spellwork'. *Feel It Break.* Domino Records, 2011.

Baudrillard, Jean. *Seduction*. New World Perspectives: Montreal, 1990.

Boyer, Anne. 'When the Lambs Rise up Against the Bird of Prey', *A Handbook of Disappointed Fate*. Ugly Duckling Presse, 2018.

Dementiev, Georgi Petrovich. *The Gyrfalcon*, 1960.

Diedrichsen, Dietrich and Anselm Francke. *The Whole Earth: California and the Disappearance of the Outside*. Sternberg Press, 2013.

Hugill, Alison. 'The instagrammable angst of Anne Imhof'. *Momus*. 5 October 2016, Online.

Jay, Mike. 'Don't Fight Sober'. *The London Review of Books*, Vol. 39 No. 1, 5 January 2017.

Macdonald, Helen. 'Military Falcons'. *Memory Maps*. University of Essex Centre for Creative Writing.

Macdonald, Helen. *H is for Hawk*. Jonathan Cape, 2014.

Obrist, Hans Ulrich. 'Choreographed Layers: Anne Imhof'. *Mousse Magazine*, Summer 2016.

Rothstein, Adam. *Drone (Object Lessons)*. Bloomsbury Academic, 2015.

Sebald, WG *Rings of Saturn*. New Directions Books, 1998.

05 Field

Wheeler, Joshua. *Acid West*. Farrar, Straus & Giroux Originals, 2018.

Press, Eyal. 'The Wounds of the Drone Warrior'. *The New York Times*, 13 June 2018.

Shenk, Joshua Wolf. 'What Makes Us Happy?' *The Atlantic*. June 2009.

06 Spirit

Feinstein, Laura. 'From Hip-Hop to High Art: Artist Korakrit Arunanondchai's Unusual Path'. *Vice*, 10 October 2013.

Freeman, Nate. 'Denim from the Drone's Eye View'. Art News. Online, 7 January 2015.

Garavi, Maryam Monalisa. 'Paranoid Androids'. *The New Inquiry*, 3 August 2012.

Haraway, Donna. *Staying with the Trouble*. Duke University Press, 2016.

LaBarge, Emily. 'Sour Heart by Jenny Zhang review – from China to the US'. *The Guardian*, 22 September 2017.

Loos, Ted. 'An Artist Who Talks Fast but Makes Meditative Films'. *The New York Times*, 27 April 2017. Online.

Morrison, Toni. 'No Place for Self-Pity, No Room for Fear'. *The Nation*, 23 March 2015.

Peckham, Robin. 'Korakrit Arunanondchai: The Denim Painter's Universe'. *Leap*, 9 December 2015. Online.

Shea, Honora. 'Korakrit Arunanondchai's Body Work'. *Interview*,

16 July 2014.

Steyerl, Hito. 'In Defense of the Poor Image'. *e-flux,* journal #10, 2009. Online.

Stone, Matthew. 'All about the boychild'. *i-d*, 13 November 2013. Online.

Timsit, Annabelle. 'What Happened at the Thailand "Black Site" Run by Trump's CIA Pick'. *The Atlantic*, 14 March 2018.

Virilio, Paul. *War and Cinema: The Logistics of Perception*. Verso Books, 2009.

Virilio, Paul. *The Vision Machine*. Indiana University Press, 1994.

07 Paradise

Iadarola, Alexander. 'Lumapit Sa Akin, Paraiso'. *dis* magazine, 2016.

Olson, Philip and Christine Labuski. 'There's always a [white] man in the loop: The gendered and racialized politics of civilian drones'. *Social Studies of Science*, Vol 48, Issue 4, 2018.

Tsui, Stephanie. 'Filipino's cancer ordeal underlines the precarious status of domestic helpers in Hong Kong', *South China Morning Post*, 23 March 2019.

08 Oculus

Bastani, Aaron. *Fully Automated Luxury Communism*. Verso Books, 2019.

Broome, Henry. 'The weird + eerie: an interview with Lawrence Lek on crossing the line + exposing the deeply embedded through VR'. *AQNB,* 13 June 2017.

Garland, Alex, *Ex Machina*, 2014.

Jonze, Spike. *Her*, 2013.

Kubrick, Stanley. *2001: A Space Odyssey*, 1968.

Lang, Iris. 'Conversation: Lawrence Lek talks Sinofuturism, automation, identity, and communism'. *Sine Theta Magazine,* 10 October 2018.

Li, Alvin. 'Lawrence Lek "Future 2065" at K11, Hong Kong'.

Mousse, May 2018.

Mamoru Oshii. *Ghost in the Shell*, 1989.

Scott, Ridley. *Blade Runner*, 1982

Tarkovsky, Andrei. *Stalker*, 1979.

Wong Kar-Wai, *2046*, 2004.

Westworld, 2016-2020.

Villeneuve, Dennis. *Blade Runner: 2049*, 2017.

Zhang, Gary Zhexi. 'Where Next?: Imagining the dawn of the "Chinese century"'. *Frieze*, 22 April 2017.

09 Simulation

Cheng, Ian. *Emissaries Guide to Worlding*. Verlag der Buchhandlung Walther Konig, 2018.

Comer, Stuart. 'Ian Cheng's Emissaries'. *MoMA Magazine*, 6 March 2019.

McLean, Matthew. 'Ian Cheng's "Emissary Forks at Perfection"'. *Art Agenda*, 17 November 2015.

Wilson, Louise. 'Cyberwar, God and Television: Interview with Paul Virilio'. *CTheory*, 1 December 1994.

CULTURE, SOCIETY & POLITICS

The modern world is at an impasse. Disasters scroll across our smartphone screens and we're invited to like, follow or upvote, but critical thinking is harder and harder to find. Rather than connecting us in common struggle and debate, the internet has sped up and deepened a long-standing process of alienation and atomization. Zer0 Books wants to work against this trend. With critical theory as our jumping off point, we aim to publish books that make our readers uncomfortable. We want to move beyond received opinions.

Zer0 Books is on the left and wants to reinvent the left. We are sick of the injustice, the suffering and the stupidity that defines both our political and cultural world, and we aim to find a new foundation for a new struggle.

If this book has helped you to clarify an idea, solve a problem or extend your knowledge, you may want to check out our online content as well. Look for Zer0 Books: Advancing Conversations in the iTunes directory and for our Zer0 Books YouTube channel.

Popular videos include:

Žižek and the Double Blackmain

The Intellectual Dark Web is a Bad Sign

Can there be an Anti-SJW Left?

Answering Jordan Peterson on Marxism

Follow us on Facebook
at https://www.facebook.com/ZeroBooks and Twitter at https://twitter.com/Zer0Books

Bestsellers from Zer0 Books include:

Give Them An Argument
Logic for the Left
Ben Burgis
Many serious leftists have learned to distrust talk of logic. This is a serious mistake.
Paperback: 978-1-78904-210-8 ebook: 978-1-78904-211-5

Poor but Sexy
Culture Clashes in Europe East and West
Agata Pyzik
How the East stayed East and the West stayed West.
Paperback: 978-1-78099-394-2 ebook: 978-1-78099-395-9

An Anthropology of Nothing in Particular
Martin Demant Frederiksen
A journey into the social lives of meaninglessness.
Paperback: 978-1-78535-699-5 ebook: 978-1-78535-700-8

In the Dust of This Planet
Horror of Philosophy vol. 1
Eugene Thacker
In the first of a series of three books on the Horror of Philosophy,
In the Dust of This Planet offers the genre of horror as a way of
thinking about the unthinkable.
Paperback: 978-1-84694-676-9 ebook: 978-1-78099-010-1

The End of Oulipo?
An Attempt to Exhaust a Movement
Lauren Elkin, Veronica Esposito
Paperback: 978-1-78099-655-4 ebook: 978-1-78099-656-1

Capitalist Realism
Is There No Alternative?
Mark Fisher
An analysis of the ways in which capitalism has presented itself
as the only realistic political-economic system.
Paperback: 978-1-84694-317-1 ebook: 978-1-78099-734-6

Rebel Rebel
Chris O'Leary
David Bowie: every single song. Everything you want to know,
everything you didn't know.
Paperback: 978-1-78099-244-0 ebook: 978-1-78099-713-1

Kill All Normies
Angela Nagle
Online culture wars from 4chan and Tumblr to Trump.
Paperback: 978-1- 78535-543-1 ebook: 978-1-78535-544-8

Cartographies of the Absolute
Alberto Toscano, Jeff Kinkle
An aesthetics of the economy for the twenty-first century.
Paperback: 978-1-78099-275-4 ebook: 978-1-78279-973-3

Malign Velocities
Accelerationism and Capitalism
Benjamin Noys
Long listed for the Bread and Roses Prize 2015, *Malign Velocities*
argues against the need for speed, tracking acceleration
as the symptom of the ongoing crises of capitalism.
Paperback: 978-1-78279-300-7 ebook: 978-1-78279-299-4

Meat Market
Female Flesh under Capitalism
Laurie Penny
A feminist dissection of women's bodies as the fleshy fulcrum of
capitalist cannibalism, whereby women are both consumers and
consumed.
Paperback: 978-1-84694-521-2 ebook: 978-1-84694-782-7

Babbling Corpse
Vaporwave and the Commodification of Ghosts
Grafton Tanner
Paperback: 978-1-78279-759-3 ebook: 978-1-78279-760-9

New Work New Culture
Work we want and a culture that strengthens us
Frithjoff Bergmann
A serious alternative for mankind and the planet.
Paperback: 978-1-78904-064-7 ebook: 978-1-78904-065-4

Romeo and Juliet in Palestine
Teaching Under Occupation
Tom Sperlinger
Life in the West Bank, the nature of pedagogy and the role of a
university under occupation.
Paperback: 978-1-78279-637-4 ebook: 978-1-78279-636-7

Ghosts of My Life
Writings on Depression, Hauntology and Lost Futures
Mark Fisher
Paperback: 978-1-78099-226-6 ebook: 978-1-78279-624-4

Sweetening the Pill
or How We Got Hooked on Hormonal Birth Control
Holly Grigg-Spall
Has contraception liberated or oppressed women?
Sweetening the Pill breaks the silence on the dark side of hormonal
contraception.
Paperback: 978-1-78099-607-3 ebook: 978-1-78099-608-0

Why Are We The Good Guys?
Reclaiming Your Mind from the Delusions of Propaganda
David Cromwell
A provocative challenge to the standard ideology that Western
power is a benevolent force in the world.
Paperback: 978-1-78099-365-2 ebook: 978-1-78099-366-9

The Writing on the Wall
On the Decomposition of Capitalism and its Critics
Anselm Jappe, Alastair Hemmens
A new approach to the meaning of social emancipation.
Paperback: 978-1-78535-581-3 ebook: 978-1-78535-582-0

Enjoying It
Candy Crush and Capitalism
Alfie Bown
A study of enjoyment and of the enjoyment of studying. Bown asks what enjoyment says about us and what we say about enjoyment, and why.
Paperback: 978-1-78535-155-6 ebook: 978-1-78535-156-3

Color, Facture, Art and Design
Iona Singh
This materialist definition of fine-art develops guidelines for architecture, design, cultural-studies and ultimately social change.
Paperback: 978-1-78099-629-5 ebook: 978-1-78099-630-1

Neglected or Misunderstood
The Radical Feminism of Shulamith Firestone
Victoria Margree
An interrogation of issues surrounding gender, biology, sexuality, work and technology, and the ways in which our imaginations continue to be in thrall to ideologies of maternity and the nuclear family.
Paperback: 978-1-78535-539-4 ebook: 978-1-78535-540-0

How to Dismantle the NHS in 10 Easy Steps (Second Edition)
Youssef El-Gingihy
The story of how your NHS was sold off and why you will have to buy private health insurance soon. A new expanded second edition with chapters on junior doctors' strikes and government blueprints for US-style healthcare.
Paperback: 978-1-78904-178-1 ebook: 978-1-78904-179-8

Digesting Recipes
The Art of Culinary Notation
Susannah Worth
A recipe is an instruction, the imperative tone of the expert, but
this constraint can offer its own kind of potential. A recipe need
not be a domestic trap but might instead offer escape – something
to fantasise about or aspire to.
Paperback: 978-1-78279-860-6 ebook: 978-1-78279-859-0

Most titles are published in paperback and as an ebook.
Paperbacks are available in traditional bookshops. Both print and
ebook formats are available online.
Follow us on Facebook
at https://www.facebook.com/ZeroBooks
and Twitter at https://twitter.com/Zer0Books